Look Up

Know Our Universe in 60 minutes
with 50 Look Up Questions and
30 Cosmic Connections

Kamal Deep Dham

Look Up
Know our Universe in 60 minutes
with 50 Look Up questions and 30 Cosmic connections

First Edition Self-Published by Kamal Deep Dham in India, 2019

All texts either quoted directly or paraphrased have been indicated by in-text citations. Full bibliographic details are given in the reference list which also contains internet sources containing URL.

For more information contact:
House Number PH -9, 4th Floor, 5th Block, Richfileds Apartment, Outer Ring Road, Opposite to Kalamandir Shop, Marathalli, Bangalore 560037

Email: kamal.dham@gmail.com

Cover Artwork and Book Art Direction by Amit Garg
Cover Photo by Greg Rakozy on Unsplash
Printed in New Delhi- India by Element94

ISBN 978-93-5351-288-0

"Look forward in space when you want to Achieve!

Look backward in time when you want to Learn!

Look Up through space-time when you want to

Explore and Contemplate!"

By Kamal Deep Dham

Table of Contents

Preface

It's only we, the Homosapiens on the Earth, who figured out hundreds of years ago, how to store information outside the human brain in form of a book. Being the evolved decedent of same species, I too had a dream to write one, on the topic which is close to my heart. Take a pause and guess what that is? Next few lines will for sure confirm if you have still not figured out by the cover page.

You might have heard many times from your parents, loved ones, teachers, elders and friends that we must always look forward and should not remain in past. From this book, I want to promote why we should instead always "**Look Up!**" Looking up gives me hope, ideas, thoughts, satisfaction and strength! It calms my neurons. It pushes me to think why we are here and for what purpose. When I look up in deep sky, I always wonder how miniscule we are, yet so important participant in the grand cosmic dance. It gives me direction, like it has out here - to pen down my passion.

Following three profound questions motivated me to write my accumulated wisdom out here:

- Who are we?
- What is the Universe?
- How are we connected to the Universe?

Who are we? Have you ever asked this question to yourself? How life started in the beginning with single cellular microbes and evolved to complex creature like us with consciousness and a brain? We are hunter, wandering species and have genes to explore. Since long we are asking questions like — why sky is blue and grass is green and exploring them logically to draw scientific answers. Step-by-step with genes of exploration we have evolved as an intelligent species on the Earth.

What is the Universe? When we look up, we see vast horizon around us! It's the same Universe which has always proved our imaginations wrong! Looking up has always thrown many questions. How Universe started? What is space?

What is time? What are stars and galaxies? Why Sun rises and sets? How Universe will end? Human kind has built technology which has answered many of these with the language of science. Many more of them are still not answered yet. Answers come and bring with them more questions and the exploration goes on!

How are we connected to Cosmos? All life on Earth is connected to the Cosmos we live in. Are there any examples? Did you ever think from where Iron (heavy element) came in our blood stream? Why do we find symmetry everywhere in the Universe? Are we made of star dust? Answers to these questions will make you think how we all are engineered in the same factory of the Universe that is driven by laws of nature!

This is my first book and I have used following three well known elements of effective writing.

- Narrate a Story: the book has story of me and the Universe.
- Ask Questions: the book has 50 Look Up questions to explain the Universe in detail.
- Find Connections: I have included 30+ cosmic connections to substantiate my belief, that how closely we are related to the Universe we live in.

I strongly believe that anyone from any age group can read it. Purpose is to promote basic understanding of Cosmos to everyone. Kids will like creative part of the Universe — like the stars, galaxies, Moon, Sun, colours and light. Youth will love science behind the Universe–the symmetry, cycles, mathematical design, and the equations. Adults will love the cosmic connections and new dimensions of space explorations — like the atomic, biological and chemical connections, quantum life, and life beyond our Earth. Elderly people will love theories that explain the reality and what is still not yet explained — like space-time, strings everywhere, what is life? What is consciousness?

First 60 pages of the book outline the story which I want to narrate. It will not take more than 60 minutes for an average reader to grasp the essence. Rest of the book is for those who find it interesting after an hour, and need more kilobytes of information to explore further!

Dedications

I never thought till few months back that I will ever write a book. Then who motivated me to do that?

Firstly, I would like to thank my Telescope, Orion Dobsonian XT10, which I bought recently. It was the first glance of bright full Moon in the eye piece of my telescope, which baffled my mind and pushed me to explore and read more to write!

Secondly, I would like to thank my wife, Richa and my twins, Navya and Nipun who gave me support in improving and reviewing the content.

Last, but not the least, my parents who have given me the birth with some genes mutated towards cosmology!

Acknowledgements

As father of Kamal, the author of this book, the news about him, having written this book, for a moment, took me by great but pleasant surprise. However, the initial glance over the pages soon turned into curious reading.

As the reading progressed, I started getting overwhelmed with a sense of pride and inspiration. The effort of the author is commendable. The presentation of the story about the Universe, interwoven with his own development story is seen as a successful attempt to make the narrative more reader friendly. Addition of ' Look Up questions ' along with their answers is a step to keep the reader's curiosity alive. We are deeply rooted into Earth and it's family of the Universe. The need to know more and more about our surroundings is natural and essential. How, when, where we got originated, where we are heading to. and more, are some very interesting facts that we all may wish to know. The fact that we are all made up of star dust and that all the Galaxies, Stars, Planets make only 4% of the Universe. Rest all is dark energy and dark matter. Isn't it quite baffling! I found the reading, a mix of bewilderment, interesting, curiosity and enrichment.

I congratulate Kamal and wish him great success in this noble venture.

Krishan Kumar Dham
Retired Mechanical Engineer from BHEL Haridwar

"Look Up" is not just the title of this book but a change in our viewpoint. Kamal discussed a lot of topics during writing of this book with me and I must admit a lot of them were astonishing facts. The most important aspect for which I say it is a viewpoint change for us is it is a very humbling experience to see a full moon, majestic rings of Saturn, latitude-like lines on Jupiter and the star factory nebulae especially in Orion constellation. This journey got us to explore the calmness of the dark sky. It reminds us about our miniscule existence when compared to the vast Universe or should I say multi-verse!

Richa Dham
BE & M-Tech in electronics and communications from IIT Delhi, Sr. Applications Manager in Cypress Semiconductors

The book "Look Up" is good insight for our Universe. At present, we are among world leaders in space exploration through ISRO. This book enlightens the young ones in the field of cosmology. Every school going and undergraduate class student must go through this book to have basic knowledge of Universe and few mathematical concepts related to astrophysics.

Subhash Thamman
Retired Professor in Physics

The book 'Look Up' not only talks about the different facts and concepts about the Universe but also justifies the reasons for the existence and connection between the life existing on Earth and the vast universe. The book also provides a set of 50 questions about Universe, which helps in further understanding and clearing of concepts talked about in this book. I am looking forward to reading this book and other books that would by written by the author and my father, Mr. Kamal Deep Dham.

Navya Dham
Higher Secondary Student in Science

The title of the book "Look Up" doesn't mean to just look at the sky and look at the celestial bodies, it means to understand how we are connected to it. What I have learned from this book is how my father's life story is connected to the cosmos. This book talks about all the concepts related to the universe out of which my favorite is the Multi-verse. Everyone who will read this book will easily learn the concepts of the universe.

Nipun Dham
Higher Secondary Student in Science

Introduction

"The Cosmos is within us, we are made of star stuff.

We are a way for the Cosmos to know itself."

By Carl Sagan

Figure: It is believed in Hindu mythology that entire Universe
resides inside the mouth of Lord Shri Krishna!

Story of Me and the Universe

There is a famous story in Indian mythology, related to Lord Shri Krishna that when he was a kid, he used to eat soil. One day, his mother Yashoda caught him and asked whether he was eating soil. Shri Krishna lied that he wasn't. To believe him, his mother asked him to open his mouth. When he opened his mouth, his mother saw the whole Universe inside, which made her go unconscious. In India, spiritually it is believed that the entire Universe is God's creation and God resides within each human being on the Earth.

Have you ever gone in to a spiritual state of mind and tried to look at the world around you with finding purpose behind everything you see, you do and you get in your lives? Mid-Age (45) is the phase of life (where I am in today!), when beings like me start contemplating life with a word "purpose". Why are we here? Who are we? Who created us? These are some doubts that push us to explore the spiritual part of the world around us. Being a passionate science (physics) freak, I never allowed my spiritual imaginations and questions to end with the answer — God's will or creation. Instead I try to explore the world with objectivity and science principles we have defined for us so far. Although, I do sometimes read metaphysics (a branch in science which deals with the first principles of things, including abstract concepts such as being, knowing, identity, time, and space.) to find connections!

Since my childhood I used to love reading about science and Universe. I am lucky I kept my passion alive, and could gather some two cents of wisdom in this book, to throw some thoughts around it. Following are three abstract questions, which motivated me to write this book.

- Cosmos: What is Cosmos (Universe)?
- We: Who are we (Human Beings)?
- Cosmic Connection: Are there any connections between us and the Cosmos we live in?

To find answers to these profound questions, I organised my acquired knowledge in ten different phases of human life. And how these phases are connected or similar with Universe we live in. These ten phases are:

1. Pre-Birth: 9 months in mother's womb
2. Birth: first day on the Earth
3. Infant: Age 1 - 5
4. Childhood: Age 6 - 12
5. Teen Age: 13 - 19
6. Youth: Age 20 - 29
7. Adulthood: Age 30 - 39
8. Mid-Age: Age 40 - 59
9. Mature Age: 60 - 80+
10. Death: Last day on the Earth

My effort here is to compare and find connections in life on Earth with the Universe we live in. Each phase of human life is outlined in separate section and each section has following three elements:

1. A story (relating to me) and the Universe
2. Explore Universe with Look Up Questions
3. Cosmic Connections

Idea is to gather information, analyse step-by-step, and find related pointers that we can reflect in the end.

Last but not the least, objective of this book is to also know the Cosmos with top **50 Look Up questions.**

Let's begin!

"When we Look Up in the deep sky, we always wonder how miniscule we are, yet an important participant in the grand cosmic dance!"

By Kamal Deep Dham

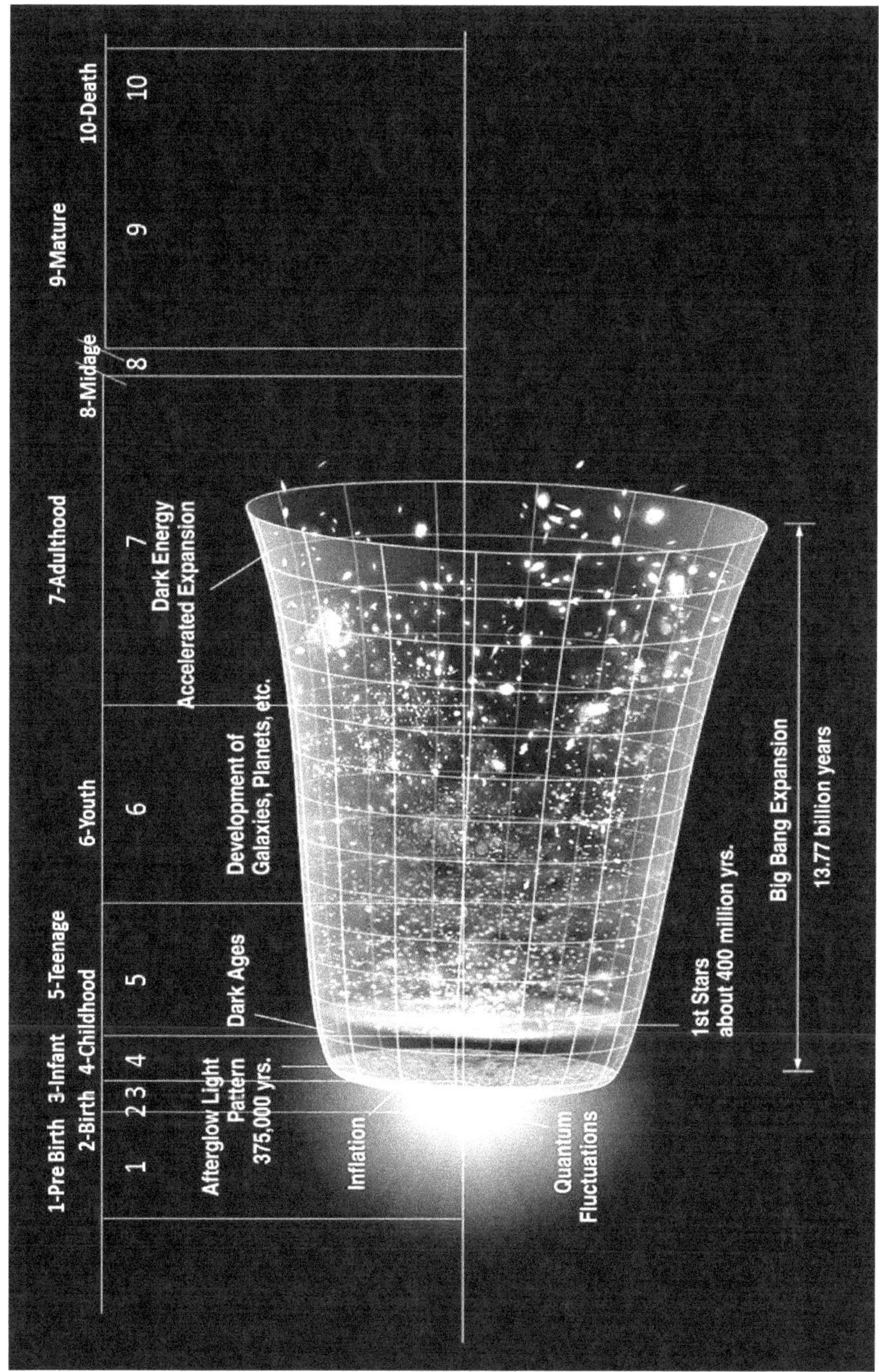

Figure: Diagram of evolution of the (observable part) Universe from the Big Bang (down) to the present.

I have marked ten phases of Universe outlined in this book.

Chapter One

Pre-birth (Care and Pain)

"There was no 'before' the beginning of our Universe,
because once upon a time, there was no time."

By John D. Barrow

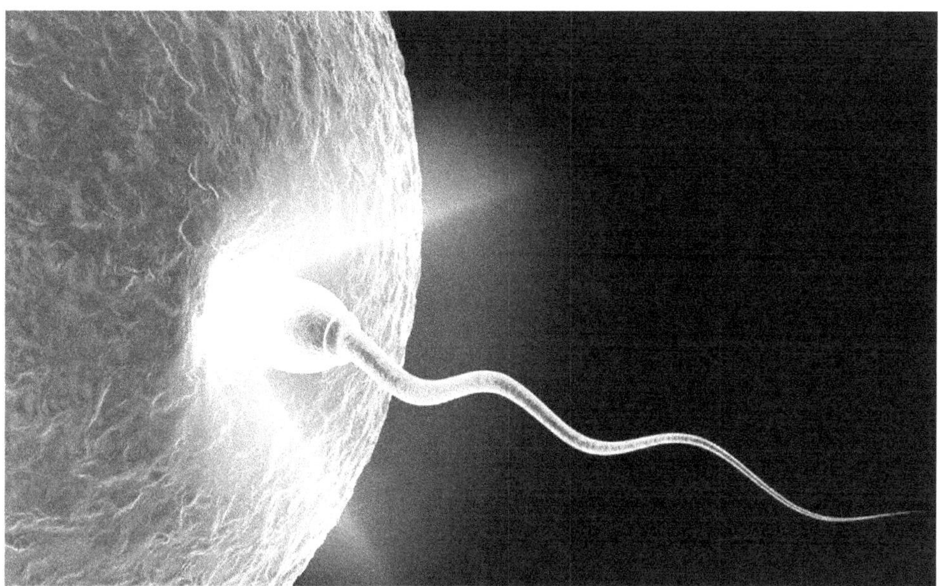

Figure: Moment before birth of new human life on Earth.

My Story!

Though my birth date is 29th Jun 1973, but probably I might have started my journey nine months before (around Oct 1972) in my mother's womb. Do you remember anything of those 40 weeks of existence inside thick uterus fluid? No one has yet claimed and provided any memory of that time.

It's totally mind-boggling but true. Within the womb of a woman it is possible to conceive and create a whole new human being!—like stars and planets are created in the nebulas of vast void of the Universe. Every single one of us is here alive today because we started growing within the womb of our mother, and we come from a long-line of ancestors born from the wombs of their mothers.

Every mother goes through 270+ days of pain, stress and restricted routine just to ensure she takes care of her baby well till the time of delivery. Being a male, I will not be able to appreciate the amount of pain and care my mother would have taken for keeping me healthy inside, till I saw first photon of light from the Sun. She would have changed her nutrition plan to ensure I get my good calories on time. She would have gone through multiple rounds of health check-up in hospital, just to ensure that my heart beat and weight graph remains proper. She would have restricted her casual movements just to ensure I don't get any jerks inside. She had patiently waited and cared for nine long months to ensure I develop all organs and immunity well to face the Earth atmosphere and gravity. The pain of joy that she experienced on 29th June 1973, I will never be able to give back till the time I finish my last breath on the Earth.

Women can use the power of their womb not just to grow babies but to tune into the creative potential of making a cohesive and progressive civilisation on the Earth. The child, who has not been born yet, could become anything—a great leader like Gandhi, Shakespeare, Einstein, Tendulkar—and thus, holds for all of humanity the principle of what we all may yet become in our lives. What I have become today? I am still trying to contemplate!

Beginning of life inside mother's womb is the most astonishing concept which mankind has to still explore and understand. We have done excellent progress in understanding the starting phase of life inside the womb. Is this pre-birth phase having any connection to the Cosmos we live in? Were there any cosmic sperm and eggs that created the Universe? Are there any similarities and hidden answers with the Universe and laws of nature we experience? Like I have my mother, does the Universe also have a care-taker? Were there any defined and restricted conditions during the pre-birth phase of the Universe? How it was conceived in the womb of Cosmos and what happened during its pregnancy period?

Story of the Universe!

As per the theory of Big Bang there was nothing before the birth of our Universe, no such thing as time or space. There was just a single hot, condensed point—(call containing all matter in the Universe we see today. In addition, all four fundamental forces (the gravitational, electromagnetic, strong and weak forces)were unified as a single force. We still don't know in detail what happened, when the Universe was in the womb of Cosmos. There are theories that suggest possible solutions, but none of them has found proven results yet. We still don't have clue on how our Universe was conceived. Many scientists are working to connect this piece of story in the big picture.

To understand the full story of the Universe, let's first make a list of queries we might have, if we relate to different phases of human life with the Cosmos around us.

I call them as **Look Up** questions out here.

Look Up Questions!

Question 1	What is the Universe? How and when the Universe began?
Question 2	What was there before the beginning (Big Bang)?
Question 3	What is space?
Question 4	What is time and when it started? What is Planck time and Planck length?
Question 5	What is actual shape of our Universe?

Instruction: *Don't Google them! I will answer all these Look Up questions in separate section, later in the book.*

How are we connected to the Cosmos? Do we see any connections relating to pre-birth phase of the Universe? You will get more details in the section named **Cosmic Connections** later in the book:

Cosmic Connections!

Connection 1	We are connected to the Universe atomically!
Connection 2	We are connected to Earth chemically!
Connection 3	Women's womb holds a powerful connection to the astronomical cycles of the Earth, the Sun and the Moon.
Connection 4	Embryonic recapitulation of all living beings on Earth, tells us how we evolved!

"Embryonic recapitulation is a signature of connection between
all lives on Earth! To know if we are living in Multiverse, we must find out
if our Universe also recapitulated in its embryonic state!"

By Kamal Deep Dham

Chapter Two

Birth (Hope)

"Because there is Law such as gravity,
the Universe can and will create itself from nothing."

By Stephen Hawking

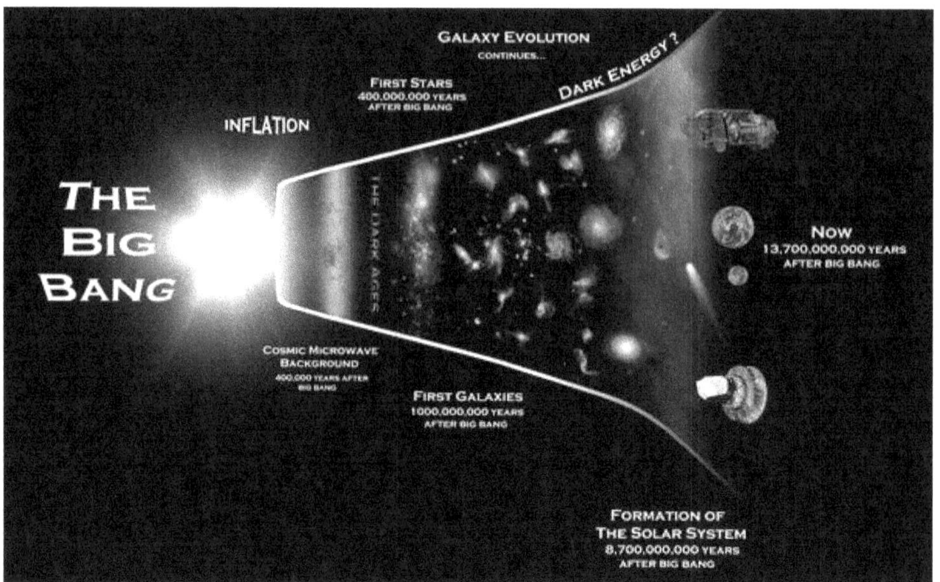

Figure: Shows the Big Bang from the left till the time now after 13.7 billion years.

My Story!

Birth of a child, brings hope in the family. Everyone in the joint family (father, mother, grandparents and many extended family members) eagerly await the birth and then participate in joint celebrations and rituals. It unites the close and dear ones with positive energy and hope of well-being in future.

I was told that my birth gave my parents the most joy of their life time. It seems I was born due to special blessings of God (*Guru Nanak DevJi*) and *Vrats* (holy fast) that my mother used to keep, offering her prayers and beliefs in the almighty. My birth was celebrated both spiritually (in *Gurudwara*) and socially. Entire quorum of Dham (My father's side) and Malik (My mother's side) families were invited in a grand event celebration.

When a child is born, it comes with a sense of optimism for the parents and the family. A sense of hope, that new life may bring something new and special into the world. In India, we have variety of rituals, attached to the new born. Most of them have spiritual connections like,

a) **First drop of honey**: As soon as the baby is born, a drop of honey is poured into his mouth and ears (just a little bit symbolically). Honey stands for sweetness. And this Hindu ritual is to ensure that the child is sweet-spoken and hears only sweet things.

b) **Naming ceremony**: On the naming ceremony or *'namkaran'* of the child, a holy fire or *'havan'* is lit up. A *'havan'* is performed to please all the Gods and then a letter from the Sanskrit alphabet is selected according to the child's *'rashi'* or Moon sign. The child's name must begin with this sacred letter so that his/her life can be very auspicious.

c) **Mundan**: The *mundan* ceremony is when the baby has his/her first haircut. According to Hindu tradition, the baby's head is shaved for the first time and the hair is offered to the Gods as a sacrifice.

There are many more you will find, based on which community you belong to in diverse India. Often you will see involvement of celestial bodies like Sun, Moon, Earth, Stars, etc. in these ritual processes. Most of the time, this connection is more spiritual, than logical. But I could find one, which seems to have objective relation,

d) **Burying the placenta**: The placenta is an organ that develops in the uterus during pregnancy, providing essential oxygen and nutrients to the growing baby. After the baby's birth, the placenta leaves the body, having fulfilled its purpose. Although most hospitals today simply dispose-of placentas after childbirth, people from different religious and cultural traditions throughout history and to this day have honoured the organ's role in nourishing the foetus. One of the rituals included burying it within the mud of Earth. I guess, this is to acknowledge that, this common organ of mother and her new born is made of same atoms which Earth is made of. Don't you think so?

Process and time involved in a birth of a human child is very well understood by us. Doctors can track from day one the growth pattern, weight, height and heartbeat of the foetus and ensure flaw less birth of a new born. We can determine the optimum date and time of the birth and sometimes choose based on the health parameters of the baby using sophisticated medical equipment.

Story of the Universe!

But do we have any knowledge and understanding how our Universe was born? What happened during its Birth? What were the building blocks of newly born Universe?

We still don't know how the Universe began but we have very strong evidence that something very interesting happed 13.7 billion years ago and that can be taken as the birth of our Universe. We call it the Big Bang.

Universe began from an unimaginably hot, dense point with a Big Bang. The Bang in the Big Bang is described, when the Universe was just 10^{-34} of a second or so old — that is, a hundredth of a billionth of a trillionth of a trillionth of a second in age. It experienced an incredible burst of expansion known as inflation, in which space itself expanded faster than the speed of light. During this period, the Universe grew doubled in size at least 90 times, going from subatomic-sized to golf-ball-sized almost instantaneously.

If you want to know more about what happened during the birth of a Universe and Cosmic Connections during this phase, please do go through the answers and explanation irrespective sections later in the book.

Look Up Questions!

Question 6	What happened during first second of the Big Bang? What were the building blocks of baby Universe?
Question 7	What is age, size and mass of the baby and current Universe?
Question 8	Is our Universe static or expanding? Can we see the whole Universe?
Question 9	Where is centre and edge of the Universe?
Question 10	What is inflation theory of the Big Bang?

Cosmic Connections!

Connection 5	We are connected to each other biologically!
Connection 6	Our blood stream has heavy element like Iron! From where it came?
Connection 7	Mutation is the cause of variety of life we see on Earth and UV light in the Universe controls mutation!

"Origin of Iron element inside the hot crust of star and

its presence in our blood stream is not a coincidence!

It's a signature of our connection to the Universe!"

By Kamal Deep Dham

Chapter Three

Infant (Vibrant)

"After Big Bang, it took less than an hour to make the atoms,
a few hundred million years to make the stars and planets,
but five billion years to make human on Earth!"

By George Gamow

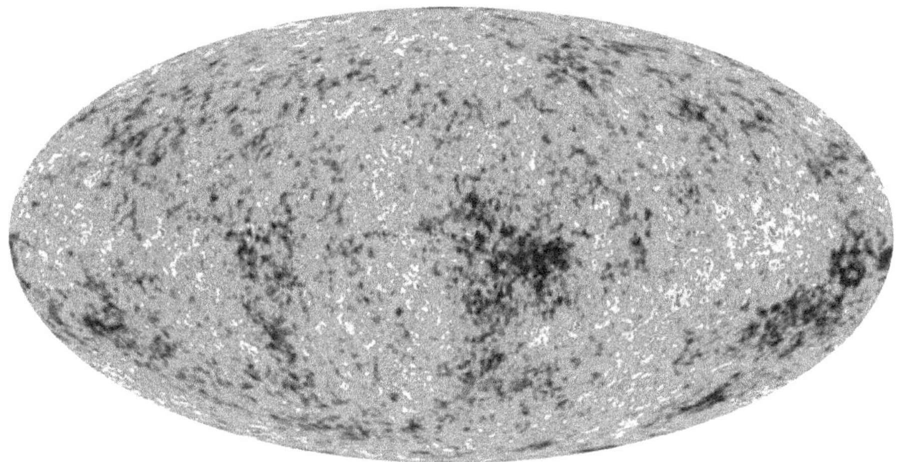

Figure: CMB, Cosmic Microwave Background, Picture of the infant Universe created from nine years of WMAP data.

My Story!

The most vibrant time for any family is when they have an infant in their home to look after. Just three kilograms of new life, makes everyone in the house more active and energetic. Why does this happen? Do you see any connection related to $E=MC^2$ law here? I guess not! It's the joy of nurturing and raising someone who has been manufactured by the same cells and atoms of the family brings in that positive energy. Growth of a new born goes through many phases till he/she jumps into a kid status.

At the time of birth, infants cannot control their body movements. Most of their movements are reflexes. Their nervous system still doesn't get fully developed. After few weeks, they start seeing clear objects which are 10 inches away from their faces. Full vision gets developed only after 6 months. Then they start crawling on the floor, and they explore the place on their hands and knees. By eight months, they can reach for and hold objects. They can pick up objects with their thumb and fore-fingers and let the objects go (drop things). When frightened, infants cry and look surprised and afraid. Infants smile in response to a pleasant sound or a full stomach. Slowly they respond when you say their name. They begin to fear strangers. They begin to fear being left by their parents. They get angry and frustrated when their needs are not met in a reasonable amount of time. Infants will talk to themselves in front of a mirror. They begin to learn what is and is not allowed. Eye contact begins to replace some of the physical contact that younger infants seek. They explore things with their mouths. They put anything they can hold, into their mouth. They respond to simple directions. By 14-18 months, many infants speak their first understandable words.

You will notice that first 2 years bring an accelerated growth in them in all forms. Physical growth is especially rapid during the first 24 months. An infant's birth weight generally doubles by 6 months and triples by the infant's first birthday. Similarly, a baby grows between 10 to 12 inches in length (or height), and the baby's proportion changes during the first 2 years. From age 2 till 5, the physical growth stabilises and toddlers learn to walk, talk and think. They become naturally mobile, active and curious. Slowly their bone structure and muscle tissues start becoming active to withhold the vibrant and naughty behaviour.

Do I remember the time when my parents taught me walking and talking? I guess no, why we cannot remember memories of first 5 years? Is this something to do with under developed neurons of the infant brain?

Story of the Universe!

Universe also followed somewhat same sequence after its birth. It inflated in the beginning (for very short fraction of a second) and then created building blocks of matter (atoms) during its initial hours. It was explosive, hot and dense during its birth and remained vibrant in its toddler age due to very high energy and pressure. We can even see snapshot of its vibrant state in radio waves spectrum today, known as CMB (Cosmic Wave Background). CMB is like a picture of baby or infant Universe. Like we don't remember our first 5 years of memories, we cannot even see (in visible spectrum) to the time when the Universe was in its infant age.

Let's get into details through **Look Up** questions of Infant Universe:

Look Up Questions!

Question 11   How baby Universe looked like? What is Cosmic Wave Background (CMB)?

Question 12   What happened after one second of the Big Bang? What was infant Universe made of? When did the first elementary particle come into existence?

Question 13   Why infant Universe was opaque?

Question 14   When Universe became transparent?

Question 15   What is gravity? Why it's considered to be a weak force (did you know this)?

Cosmic Connections!

Connection 8   Every time we take a breath, we take atoms from the breath of other living beings ever lived on the Earth.

Connection 9   There are only six degrees of separation between us to any human on Earth!

Connection 10   There are around 100 trillion microbes/bacteria living in our body. These microbes were the first form of life that originated on the Earth!

"By fixing the speed of light, Universe allows us to look into its past!"

By Kamal Deep Dham

Chapter Four

Childhood (Dream and Play)

"You will never find rainbows when you are looking down!"

By Charlie Chaplin

Figure: Timeline of the early Universe, showing the Dark Ages and
the First Light - when stars began to appear, causing re-ionization.

My Story!

I vaguely remember, I was probably in 4th standard! Me and my 2 little sisters used to love Saturday nights. The reason was my father used to allow us to sleep on our house terrace roof once a week. Yes, in many parts of India it is possible to sleep (especially in summers) under the dark open sky. I used to love those nights, not only because we used to get chance to sleep as a family together in a row, but also used to get IMAX view of the full sky from the top. It used to take around 20 minutes for me to get into sleep state and the easy way to sleep was to look up and watch the bright members of the dark sky!

My father used to often show us the Pole star and *Saptrishi* constellation (The *Ursa* Major and *Ursa* Minor). We also used to track other different constellations of stars every night. Probably regular habit of looking up (in such a young age) made some strange positive connections in my brain about the Universe. For me this positive connection from childhood has reignited the look up quest, at the age of 45.

Childhood comes with a passion of play and I was not any different kind of soul. Thank God! At that time there were no smartphones, iPad and online connected community. I was small-town kid, who used to love playing out in the ground or on the roads in front of my house. That night I remember I was probably 9 yrs old, and it was around 10:30 pm when my mother came into my room and found me still awake. I was lying on the bed and taking rolls from one corner to the other. She looked at me and said, she knows what's going in my mind and explained me as to why I should sleep now so that I can enjoy the whole next day! It was the excitement and anxiety in my brain that was keeping me restless and awake. I was almost counting each minute and hour in the clock which was hanging just in front of my bed. Probably, I was praying to the almighty to change the laws of physics for some time and turn the rotation of the Earth a little faster that night. I was desperately waiting for the Sun to show up its face through the small window of my room.

The next day was the most awaited day for a typical boy like me in the small town of north India. The day when I used to get opportunity to keep looking up in the sky for almost the whole day! It is the day when you can see beautiful colours in the sky. I didn't mean rainbow here! It used to be different colours of papers flying all over the blue horizon, each controlled by a string in the hands of skilled young lad like me! Yes, the day was Basant Panchami (kite flying festival)! Kite flying is very popular in many parts of India, and each region of India has one day which is celebrated as kite flying day. It used to be Basant Panchami (somewhere in February every year) at the place where I grew up and did my schooling.

You can see lots and lots of kites of different shapes, sizes and colours almost throughout the day. I have never realized that though flying kites need a special skill, but looking up in the sky for the whole day probably did not. It was that day when I loved the Sun and used to hate seeing bright and beautiful Moon coming up slowly in dark sky, signalling end of the day! The Sun and the Moon used to be my natural clocks for that full day of joy! When you watch dark sky before sleep or fly kite during the day, you by design need to look up most of the time. This makes you indirectly and unknowingly know a lot about laws of nature and what is up in the sky. Small age brings lots and lots of basic questions in your mind around what's up there.

Story of the Universe!

The Childhood phase of the Universe that I have mapped here in the book, started just after the Inflation and lasted till first set of stars were born and the first light originated. The period lasted for around 560 million long years. We still don't know in detail what happened in this phase and why most of this was Dark period when Universe was in dreaming state! It is called Dark phase, because stars didn't come into existence. Due to this for millions of years, there was no new light being produced. The dark-age only came to an end when the first stars appeared and their light started to propagate in the space. These first stars were very big ones, with some being as large as 300 times the mass of the Sun.

Let's get into **Look Up** questions around Stars!

Look Up Questions!

Question 16	How stars are formed and how they die? When will our Sun die?
Question 17	What are different types of stars?
Question 18	What are nebulae and how they are formed? What are different types of nebulae?
Question 19	Why some stars twinkle and some don't? How far is the closest star from the Earth?
Question 20	What are constellations? What are star clusters and binary star system?

Cosmic Connections!

Connection 11	Human brain mirrors visual phenomenon in the natural Universe

"We are still exploring notion of time and the Universe has
natural clocks everywhere!"

By Kamal Deep Dham

Chapter Five

Teenage (Study and Peace)

*"Good thing about Science & Math is that
it's true whether you believe in it or not!"*

By Neil DeGrasse Tyson

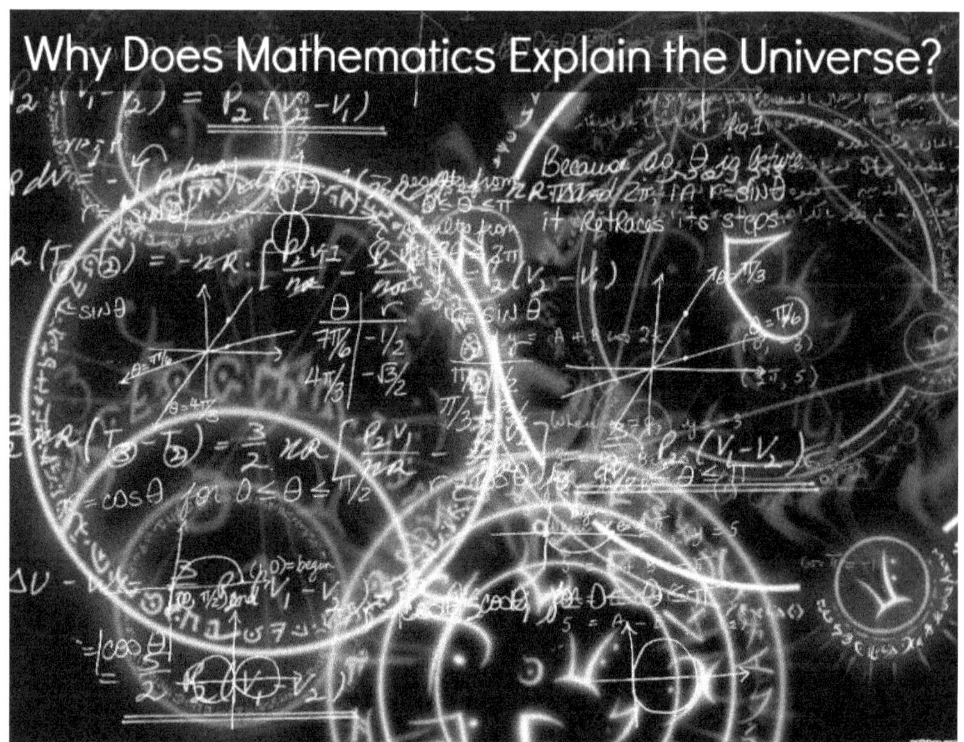

Figure: Mathematics is the language of Universe!

My Story!

Pressure of studies start during your teen age! You get constant directives from your parents on how much more marks you should get! You get into peer pressure of performance with your classmates. Especially this push increases multi-fold when your own siblings are doing comparatively better than you! I too used to strain my neurons often because of it. Somehow my brain never liked the idea of storing knowledge, just for the sake of it! I used to hate subjects which needed lots and lots of RAM bytes to store information for the examination days. I don't want to name these subjects, but will become obvious if you read my full book.

My favourite subjects used to be Science (Physics) and Math. I used to like the fact that whatever I read, I can practice that too in my home with my self collected hobby gears andtools. I remember when I made my first door bell (not with electronics) with self-made copper wire coil. It was moment of joy when I learnt about motors/dynamo and then used one (from my old toy) in my bicycle rim to light some LEDs. I used to be an In-demand kid on Diwali nights (festival of lights and crackers) due to the skill I got of troubleshooting and fixing Diwali lights with ease. If you really want to understand fundamentals of electrical circuits, you must learn how to trouble shoot and fix these beautiful light series! This was the time of my teen age, when I developed the interest to know physical laws of nature and our Universe. Force of gravity used to be my favourite, not because of Newton's apple, but for the fact that this force used to be the one that I had to compete within making my kite fly.

The science which we (mankind) have developed so far, has clear answers of almost all the **Look Up** questions that a high school teen could have thought. I remember I read tons of material on solar eclipse in my 8th Standard as we were going to witness one partial one soon during that summer. It used to amaze me how Moon sometimes can out shadow the glow of the mighty Sun.

Teen age comes with many changes in our lives too! Your brain gets instructions from many dimensions to improve and show results. Your parents, siblings put pressure for studies and to perform academically. You get pressure of your overall looks and personality in the open society. You often find yourself stressed between your physical and emotional changes. These all vectors build huge pressure on the teeny brain!

What I used to do when I was in state of being pushed? I used to like my study table, because it was adjacent to the small window of my room from where I could see big slice of open sky. I just used to look up in the sky. Whenever stressed, looking up, used to have a soothing effect on me. I am not sure why, but still today when I look up, I get calm in few minutes. The circuits of my brain get noise-free and then get stabilized.

Story of the Universe!

Teen age is also the age when we like to make friends of same school, same locality and with same hobbies. We prefer to become part of group and start linking bonding activities. During teen age of the Universe, Galaxies started to come into existence. Galaxies are formed by bonding clusters of billions of stars. Based on recent astronomical observations, formation of initial Galaxies might have come into existence as early as 200 to 600 million years after the Big Bang.

Though our laws of Physics came into existence as early as initial seconds of the Big Bang, but I have mapped it here in the teen age just to relate to my story! I developed my passion and love towards it when I was in my teen age!

Let's find out more details around galaxies and laws of the nature through **Look Up** questions of teeny Universe:

Look Up Questions!

Question 21	What are galaxies and galaxy clusters? How are they formed?
Question 22	How many types and numbers of galaxies are there in the Universe?
Question 23	What is the Milky Way and how big it is? When was it born? Why can't we see its spiral shape from our Earth like we can see Andromeda? Where is our Earth located in its spiral design?
Question 24	Why galaxies collide in expanding Universe? When will Milky Way merge with Andromeda?
Question 25	Why we find super massive blackholes at the centre of many old and large galaxies?

Just read all these questions carefully! Don't you see a pattern? These questions for sure motivate us to explore!

Cosmic Connections related to teenage have a lot to do with Science and Math!

Cosmic Connections!

Connection 12	Mathematics is the language of the Universe!
Connection 13	Fibonacci Series and Golden Ratio is everywhere in the design of Universe!
Connection 14	Pi is not just a collection of random digits. Pi is a journey!
Connection 15	Prime Numbers! It was all so neat and elegant. Numbers that refuse to cooperate, that don't change or divide, numbers that remain themselves for all eternity.
Connection 16	Power of Symmetry: Nature and Universe both loves symmetry!

"Science will never die, it only consolidates and evolves itself

till another genius like Einstein comes and mutates its genome!"

By Kamal Deep Dham

Chapter Six

Youth (Imagination and Love)

"Imagination is more important than knowledge!"

"You can't blame Gravity for falling in Love!"

By Albert Einstein

Figure: Milky Way galaxy plane seen from the Earth.

My Story!

Teen tag ends at 20 and you find yourself in the youth age! The age group that comes with imaginations and first feeling of the four-letter word **LOVE**!

In India, after completing 12th the key focus of young guns gets diverted to ensuring from where your bread and butter would come in future. What should I do next so that I get the best paying options in future? Many times, we ignore our own imaginations and passion! We don't listen to our heart and go normally with what is selling more in the market. Luckily, I made a balanced choice! I always wanted something that was based on science, technology and engineering. I used to love reading about the space and the Universe!

I joined an engineering college with one of my favourite stream i.e. computer science.

My engineering college (*GBPEC Pauri @ http://www.gbpec.ac.in*) is located at the mountain top, in the beautiful Himalayan ranges. To be precise, its location is at 30.15°N 78.78°E. It is located 1,814 meters (5,951feet) above the sea level. I used to love that place and enjoyed each day of my four-year engineering course. Though the life on mountains is normally tough, but it provides great imagination space for exploring the young minds. Tons of engineering lecture hours every day and being in the computer lab for hours and hours became part of my life like a typical engineer graduate life. I used to love computer lab sessions. I remember when I made my first digital clock software fully coded in BASIC language! Anyone remembers this first interpreter that came with X86 series of Personal Computers? Youth of today might not have heard about it even in their books. We, gang of three (Tri-Gang) mates used to sit for hours and hours in our computer lab to code, test and run (in typical language of a software freaks!). We used to enjoy breaks in night time the most, within our busy lab hours. When you sit under the dark sky and see the vastness of space from top Mountain View, you get to see all together a different and exuberant view of lifetime. Believe me, on a clear night on a mountain, you will find totally a different view of sky what you get normally in polluted (do you know light pollution?) metro cities of India. You get to see many times more stars, brighter Moon and glimpse of Milky Way plane! Have you ever seen the Milky Way? It is breath taking for sure, when you happen to see it the first time. Look at the image at start of this chapter.

We develop problem solving skills when we are working to solve tough issues. If you know software and programming language, you will appreciate the trouble shooting skills in ensuring a program runs. Many times, you get stuck for long to make a piece of code, work properly. For us (Tri gang) the most frustrating time used to be when we were not able to crack what's wrong in the code. Sometimes you start believing that it's not my program that has issue, it's probably the compiler (It used to be C language) which has some strange bug! Or there must be some virus which is creating trouble in our trouble shooting. It was then that we used to take our night breaks. We three used to go out and sit under the dark sky for some time with a hot cup of tea. When you sit on top of a mountain under a dark sky, you get to see only space and the Universe wherever you turn your head!

Those moments of star gazing, used to shift our brain gear into imagination mode. Suddenly you forget the loops, if, else and return statements of C language, in which you were drenched! Your mind gets clear with soothing breeze of imaginations; you end up with new ideas. Most of the time these breaks used to bring success in making our software program work. Take a Break! for sure was one of the important lessons during the college life that I learnt for my lifetime!

In this age of 20s, I don't know what and why it happens, but you start liking English and Hindi words which relate to feeling of LOVE! You start liking poetry, songs, shayari, romantic movies, quotes etc. And if you go through any of this literature material, you will notice that use of heavenly bodies is very common in the illustrations. To give you some example quotes:

- As long as we are under the same Moon, I feel close to you!

- I love you to the Moon and back!

- If you will be my star, I will be your sky!

- If a star fell each time I thought of you, the sky would be empty!

- Love is space and time measured by heart!

- So, I love you because the entire Universe conspired to help me find you!

- Infinite Universe, Solar System, 9 Planets, 204 countries, 809 Islands, 7 Seas, 7.4 billion people, And I am still Single!

Suddenly you relate the Moon to the person whom you are attracted to. You sometimes ditch your Tri gang too and go out alone in the dark! Watch all those twinkling stars for a while, in anticipation that you will get to see someone special soon (at least in your dreams). Well did it happen to me? I will not say no, but you know in hostel life, your friends ensure to make you imagine, that there is someone special somewhere in the Universe for all, waiting to be discovered. Well let me stop this topic! I lack the experience for this topic you know! Point which I wanted to make, was that it's the LOVE and imagination in youth, that make us Look Up often.

Figure: Our Solar system and its 8 planets.

Story of the Universe!

Out here, I have mapped youth age of the Universe to the period when our solar system was getting born. Our solar system was formed 4.6 billion years ago. Scientists believe that the solar system was formed in Milky Way, when a cloud of gas and dust in space was disturbed, maybe by the explosion of a nearby star (called a supernova). This explosion made waves in space that squeezed the cloud of gas and dust. Squeezing made the cloud start to collapse, as gravity pulled the gas and dust together, forming a solar nebula. Just like a dancer who spins faster as she pulls in her arms, the cloud began to spin as it collapsed.

Eventually, the cloud grew hotter and denser in the centre, with a disk of gas and dust surrounding it that was hot in the centre but cool at the edges. As the disk got thinner and thinner, particles began to stick together and form clumps. Some clumps got bigger, as particles and small clumps stuck to them, eventually forming planets or Moons. Near the centre of the cloud, where planets like Earth formed, only rocky material could stand the great heat. Icy matter settled in the outer regions of the disk along with rocky material, where the giant planets like Jupiter formed.

As the cloud continued to fall in, the centre eventually got so hot that it became a star, the Sun, and blew most of the gas and dust of the new solar system with a strong stellar wind. By studying meteorites, which are thought to be left over from this early phase of the solar system, scientists have found that the solar system is about 4.6 billion years old!

Let's see top **Look Up questions** on solar system and find answers in more details:

Look Up Questions!

Question 26 How and when our solar system was formed?

Question 27 How and when Moon was born? What are its building blocks?

Question 28 What are black spots on the Moon and on the Sun?

Question 29 Why we only one side of the Moon from the Earth (did you know this?)?

Question 30 Curious questions on our planets:

    a) What is Mar's atmosphere made of?

    b) Why Saturn has rings but other planets of the solar system doesn't?

    c) Why Jupiter is considered as hot gas giant? Is Jupiter a failed star?

    d) Pluto was well recognized last planet of the solar system, why its planet status is removed now?

Cosmic Connections!

Connection 17 Periodic Table symmetry: The element of surprise was not allowed near the Periodic Table!

Connection 18 Everything Spins around the Nucleus and in an orbit!

Connection 19 Cycles everywhere: Universe will keep hitting you with same lessons, until you learn!

Connection 20 Water has memory: All water has perfect memory and is forever trying to get back to where it was!

"Look forward in space when you want to achieve!

Look backward in time when you want to learn!

Look Up through space-time when you want to explore and contemplate!"

By Kamal Deep Dham

Chapter Seven

Adulthood (Space and Time)

*"Time and space are modes by which we think and
not conditions in which we live!"*

By Albert Einstein

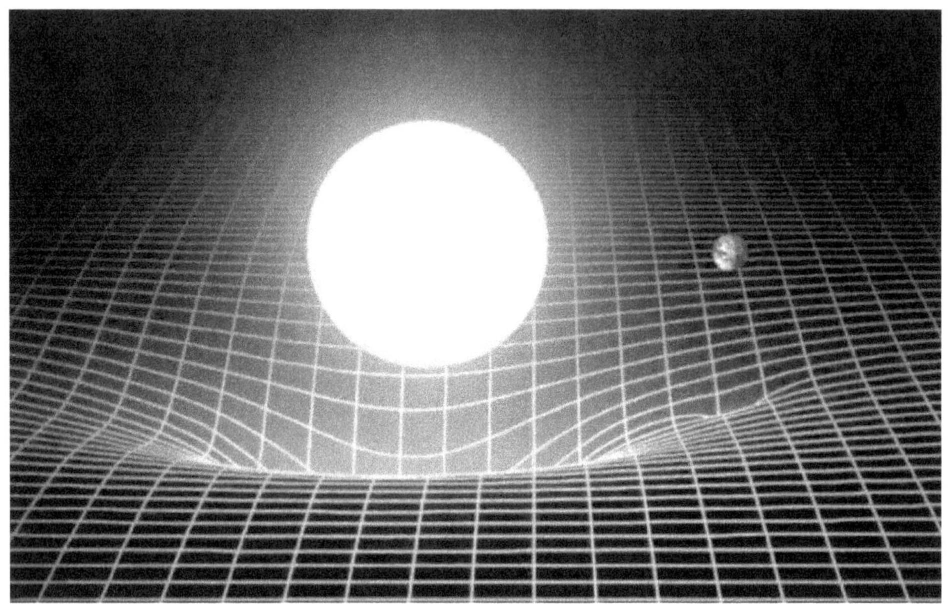

Figure: Bending of space-time fabric by Sun.

My Story!

Engineering degree in computer science for sure helps you in finding suitable job in Indian dominated IT sector. I too did not face any big trouble to find where my bread and butter would be. After my graduation in engineering, I came to Bangalore (Indian Silicon Valley) and started job in one of the well-known Indian IT giant. Life looks so beautiful and easy when you start your bachelorhood. With income in hand and tons of ambitions in your brain (developed during college days), you start moving ahead very fast in life. You find lots and lots of space and time to work on your goals and dreams. You want to prove to the world that you have the best talent in the Universe and one-day people around you will wait in queue to take an autograph from you. I call it passion at its peak.

I too went through this time and used to see dreams of changing the world (if not the Universe)! I used to have enough space and time to work hard towards my professional life as a software engineer and also explore my hobbies that I used to love. I used to love reading and watching movies about Universe. I remember during my bachelorhood, I did see many movies like CONTACT, APOLLO 13, ARMAGEDDON, STARTREK, etc. Till today I never miss any opportunity, recently I liked INTERSTELLAR, MARTIAN, GRAVITY the most. I also used to read about astrophysics in my free time. I am a big fan of the man who changed the meaning of Space and Time through his theory of relativity, you guessed it right – he is none other than legendary Elbert Einstein.

Though this theory was produced in early 19th century, but I got hang of it only when I started reading in detail during my adulthood. I used to get amazed that how can time dimension be relative? How come gravity is not a real force, but an outcome of warp in space and time fabric? Through his theory, Einstein predicted (100 years back) many mind-boggling phenomenon. We still are in process of experimenting and proving them to be correct. Some of them are:

- Idea that Universe is running on a plane which is nothing but Space and Time fabric

- Space and Time fabric can have holes too, which he called black holes

- Gravitational waves in the fabric of Space and Time. (Recently detected at LIGO lab in USA.)

- Gravity can bend light too.

- Speed of light is not an idea, it's a law.

- Gravity in which you stay and Speed at which you travel, can slow down the time with respect to reference.

Even today whenever I get a chance I do read or watch videos about him and his life. Contrary to Einstein's theory of relativity, somehow time flies much faster when you are in your bachelor life. Somewhere on fast forward mode,

TWO critical moments of change happen to most of us (especially in India), I guess. Its marriage and then comes parenthood.

Though every guy in India tries his best to find suitable life partner on his own, but most of them end up in settling with a well-defined process of Indian arranged marriage. I too did try my luck in the open sea, but then it was my father who found the best gift of my lifetime. I got married in year 2001 when I was 29 and well settled in my job. The only two conditions that I wanted to draw during the arranged marriage phase were,

a) I wanted to marry a working companion who can stand by her own and

b) Who will not have any issues living in a joint family (along with my parents) after marriage

I still say to my wife often that I am the luckiest person on Earth who has got a partner like her. Marriage brings a huge change in life. It comes with demand of commitment, ownership, care and attention. And these elements of demand consume major slice on the 24 hours' time pie chart. You suddenly start thinking as to what happened to the time? Why can't we buy few hours from market that we can spend on things which we love? Why everyday ends up with backlog of things to be completed for the next day? You start reading about Time Management, in the anticipation that you will be able to slowdown the time slice of each tick of second's hand in the clock. It slowly starts building pressure of what I call – "find time syndrome".

You start de-prioritizing occasional activities. You don't get enough time to think about yourself and your passion. You start spending time mostly in your office and then back at home on the huge list of To-Dos! Let me ask you, in that busy state, when your wife is not at home or you are alone, what would you normally do? Well I go to my open front terrace and just Look Up for some time. Have been doing this regularly, even together with my wife when we both need a break from Space and Time fabric! I believe when you make your busy brain look up, it somehow connects with Space and Time fabric and slows down the delusional effect of rat race. It gives you directions in which we should focus our energies.

Let me now talk about most interesting phase of my life before I write down Look Up questions of adult hood! It was July 2002, and I was deputed in Belgium for a 3 months official assignment through my company. Normally onsite assignments for a software engineer are like a dream come true! But this one was a bit tough on me. I left my wife back in India during her pregnancy (of twins!) period, the time when partner needs you the most. I was counting every day and used to pray almighty to speed up some time in my wrist watch! I remember, on 42nd day when I was in office (at our customer premises) at around 2 PM Belgium time, I was in meeting room with our customer and then my phone rang. It was a call from my father. I didn't pick it for the first time, but then when it rang again, I took apology from the session and went out of the meeting room to take the call. Even before I could ask the reason of call, my father said, Kamal, you must travel back soon, as soon as possible, during the same week. I was a bit confused and worried for this sudden recall from my family back in India. I spoke to my wife in detail and then realized that she was advised to get admitted in the hospital for the preparation of premature twin's delivery. I just asked her, how come in just 7 months? She said, how can she control if the two little angels don't want to stay inside her womb any longer! It was mixed feelings for me and my wife, on one hand we both were happy to kick start our parenthood phase, but on the other hand we were worried how things will go during premature birth. With no more thoughts in my mind, I travelled back the very next day and joined my wife back in India. My travel back helped us emotionally and God had given some more days for my twins to stay back attached with the umbilical cord inside their mom's womb! Finally, I was blessed with healthy twins (a girl and a boy) on 23rd Sep 2003.

Kids occupy very small space in mother's belly even when they grow from single cell to multi-cellular. In case of twins, they somehow learn and adjust in the small area wrapped and cushioned around each other. That tiny space inside mother's womb becomes their home for full nine months. But just after birth, I don't know what happens to these 9-month lesson of space adjustments! New born comes with joy, happiness and strength in everyone's life. But it also comes with a dire need of more time along with double demand of space! And as we are talking about twins, then you all can understand what folds of time and squares of space we would need. Suddenly we both became the busiest couple in the town. Our home started looking small with increment of two little head counts. Hours started running fast and definition of sound sleep changed. Time and space both run in opposite direction when you have growing twins! (Time runs fast and space starts contracting). You must have heard a quote that became our motto - "No one can scare us, we have Twins!" There are some moments in life when you become spiritual and start asking questions that no one could answer so far through science. None of the science have cracked how life started and complex cellular beings like us became conscious? Who is running the entire Cosmos where we have trillions of cubic space and dust?

Story of the Universe!

Based on mapping of age out here in this book, at present we are in the adulthood phase of the Universe. It is the time when our solar system was born and till today when the complex life like us walks on the surface of the planet Earth. Life evolution from single cell to complex organisms took long time and is a separate subject in its own. As I don't have much knowledge about the same, I will outline mainly the discoveries mankind have made in past two centuries in cosmology. Imagination of the Universe by human beings started way back in our history. It was during the 4th century BC, that Plato and Aristotle created explanation on the geocentric Universe that today secures its place as the predominant cosmological theory. According to Plato, the Earth was a sphere and the stationary centre of the Universe. The stars and planets were carried around the Earth on spheres or circles, arranged in the order of distance from the centre. These were the Moon, the Sun, Venus, Mercury, Mars, Jupiter, Saturn, and the fixed stars. Then during 16th century, Nicolas Copernicus proposed heliocentric model that proposed Sun at the centre of the Universe, with the planets (including Earth) and stars orbiting it. Till 1920s it was generally believed that there were no galaxies other than our own Milky Way and our solar system is part of it. Later we discovered that our Universe is made up of billions and billions of galaxies. This is the proven picture of Universe we have in current cosmology era.

Till today, many smart individuals walked the surface of the Earth who has greatly contributed in defining the laws of nature and mysteries of Universe we live in. Key Examples are:

1514	Heliocentrism: Nicholas Copernicus
1589	Galileo's Leaning Tower of Pisa experiment: Galileo Galilei
1687	Laws of Motion and Law of Gravitation and calculus: Isaac Newton
1782	Conservation of matter: Lavoisier
1803	Atomic theory of matter: John Dalton
1814	Wave theory of light, interference: Fresnel
1831	Electromagnetic induction: Michael Faraday
1843	Conservation of energy: Julius Robert von Mayer, William Thomson, 1st Baron Kelvin
1863	Entropy: Rudolf Clausius
1867	Dynamic Theory of Gases, James Clerk Maxwell
1887	Electromagnetic Waves: Heinrich Rudolf Hertz
1895	X-Rays discovered: Wilhelm Röntgen
1896	Radioactivity: Henri Becquerel
1897	Electron discovered: J. J. Thomson

1905	Special Relativity: Albert Einstein
	Photoelectric Effect: Albert Einstein
	Brownian motion: Albert Einstein
1913	Bohr Model of the atom: Niels Bohr
1916	General Relativity: Albert Einstein
1926	Schrödinger Equation: Erwin Schrödinger
1927	Big Bang: Georges Lemaitre
1927	Uncertainty Principle: Werner Heisenberg
1928	Antimatter predicted: Paul Dirac
1929	Expansion of the Universe Confirmed: Edwin Hubble
1932	Antimatter discovered: Carl David Anderson
	Neutron discovered: James Chadwick
1981	Theory of cosmic inflation
	Fractional quantum Hall effect discovered
1984	W and Z bosons directly observed
1998	Accelerating Universe discovered
1998	Atmospheric neutrino oscillation established
2000	Tau neutrino discovered
2012	Higgs Boson discovered
2015	Gravitational waves detected

Each discovery is a long story in itself and changed the way we think and perceive our place in the Universe.

I will mainly focus on **Look Up questions** around space and time as Einstein envisioned!

Look Up Questions!

Question 31	What is Einstein theory of relativity and how it had redefined the laws of Universe?
Question 32	Why speed of light is always constant?
Question 33	What are gravitational waves and how they are detected?
Question 34	What are black holes? How they are formed and how we detect them?
Question 35	What is a concept called wormholes, do they really exist?

Cosmic Connections!

Connection 21	Moon creates tides in ocean and Sun controls the seasons on Earth!
Connection 22	Why Plants/flowers face towards Sunlight?
Connection 23	Biological Clock is tuned to 24 hours on Earth!
Connection 24	How birds can navigate long distances? Does Earth magnetic field help them?
Connection 25	All skin colours, whether light or dark, are not due to race but to adaptation of life under the Sun!

"For Interstellar travel, it's not the space we have to navigate;

it's the time we have to vanquish!"

By Kamal Deep Dham

Chapter Eight

Mid Age (Contemplation)

"The Universe is under no obligation to make sense to you!"

By Neil deGrasse Tyson

Figure: Hubble Ultra-Deep Field image (full range of ultraviolet to near-infrared light) includes some of the most distant galaxies to have been imaged by an optical telescope, existing shortly after the Big Bang (June 2014).

My Story!

In mid age, which I am in today, we often look back and analyze what all happened to us. We contemplate various aspects of our life and start searching answers which make sense. We look for answers that should tell the sensible story and solve the problems we might have faced. We search for connections which we might have missed in understanding the bigger picture. To get into this kind of scan-mode, we often need peaceful time and space where we can examine the past events to draw sensible answers. We often need to turn our brain into a zone when it can ponder better and clear. I call this state as brain-high zone. There are many ways to get into brain-high state. Some of us use "Yoga" or "Pranayama" to get there. Many use music. Many of us get there by dance, sports and various hobbies we like. Most of the time pursuing your passion turns the brain-high state on. What I do? I pursue by following two passions that I have:

- I go to my terrace with my telescope in the dark night and start looking up in the deep sky

- Apart from cosmology, I am also passionate about long distance endurance sports. I love to run, cycle and swim long distances.

Both passions kick off my brain-high zone. In running we often call this state as runners high. This state gives me clear mind and positive energy to contemplate. I often get ideas, direction and answers to the questions I am trying to resolve. Let me give an example from my endurance journey. I got hit by an acute injury and then how I fixed it with power of deep contemplation.

In 2015 on Valentine's Day I woke up early in the morning at 5 AM and started my planned 15 kilo meters run on the road track near my home. After 6 kilometers of run, I suddenly had weird feeling in my head that urged me to slow down, though I did not stop immediately. After a km when I felt the increase in gravity of dizziness, I stopped and sat on one side of the track for 5 minutes. Took a sip of my energy drink and felt okay. I did not realize that my decision of running again after that will cost me one full year of recovery time. At around 9 kilo meters mark, I was hit with acute Vertigo attack. Everything around me started spinning and I lost my balance. I somehow again sat on side of the road and realized that the whole world is spinning around me as if I am at the centre of Universe. I was not even able to get up to look for help.

Somehow, I called my wife and waited for 30 minutes in same semi-conscious state on the road. I felt first time that my time on Earth may be over that day. After 30 minutes, I saw my wife holding me up and pushing me in one of the Auto (Indian 3-wheeler taxi) with the help of the auto driver. I reached home in the same state and vomited gallons of running drink that I took on the track. After sleeping for couple of hours I felt okay and thought that it might have happened due to some stomach imbalance or may be due to salt reduction in my body. Next morning as soon as I got up to go to the restroom, I suddenly got an even severe spinning attack. My parents and wife immediately took me to the hospital where I was kept in the emergency room. Doctors first checked the heart status and the blood parameters, when they saw that my BP was normal and there is no issue in blood reports, they did the CT scan and concluded that it was a Vertigo attack.

In Vertigo, your brain somehow gets wrong signals of the 3 dimensions around you and that makes your balancing actions totally confused. My fourth dimension (time) sense was okay though! I was advised to consult neurologist for the immediate treatment. I remember my first meeting with the consultant shattered me when he told me that I have to stop thinking about running again. Though I was barely able to walk by myself, but still I was not able to digest that I will not be able to run again. My recovery from vertigo started, it was very slow and emotionally painful. I was off from work for a full month and then worked from home for another 2 months. I was advised to not even drive. For walking a little longer distance also, I required help from my wife to hold my hand. Acute headache made me sleepless many times during the night. I even tried sleeping pills to get some sound sleep during those days. After around 4 months I joined office again. This inability to continue my running passion was taking emotional toll on me.

Then came the contemplation phase; during this forced injury break with a will to not to give up on my passion, I decided to look back and find the solution to run again. I started reading about sports injury and the importance of physiotherapy in recovery. I started physio sessions with my fitness instructor and realised that I am getting better. Many hours of balancing exercises, core and neck strengthening exercises helped me in regaining my close to normal life again. But running again was a big battle that I fought between my heart and the brain. My mind was instructed by my doctor to stop thinking about running and my heart was hardwired to put the running shoes again. By this time, I have learnt two important lessons of my endurance journey. First one - our body is the best machine ever built in the Universe, it auto corrects itself if you make your brain (mind) agree to it. You must negotiate with your brain that you can bounce back and start the process from scratch with baby steps and trust me it can be a hard nut to crack! Did the same, one Sunday, I woke up in the morning, put my shoes on and started running. I was able to run 500 meters and after that when I felt a heavy head, I stopped. After two days, again I did the same and could do 700 meters and then stopped.

Slowly in three months I could run 5 kilo meters again in 40 minutes (my earlier starting point). The lesson I learnt was that do not be tempted to increase in bursts but follow a 5% increase in 2 weeks rule, and it worked for me. Second lesson is, we must listen to our body carefully and limit the efforts accordingly. Body sends signals to our brain when it is getting over stretched, over loaded with food, when to take what kind of food, when to take day off from workout etc. You must learn how to read these signals and respect the same. Cause of all sports injuries is not appreciating the voice of your body.

This type of successful recovery makes you more mature in the subject and increase plasticity of your brain towards contemplation. Your search of logical connections that make sense, build your perspective towards life. May be somewhat same is happening to me today in my 40s and that is driving me to pen down these thoughts out here!

Story of the Universe!

Let's switch our flow towards Cosmos! We have seen the journey till our solar system and what key discoveries happed in cosmology. 21st century is considered as golden era of cosmology. That's really true. Cosmology at this present time is undergoing a transition from being a bunch of speculations to being a genuine branch of hard science, where theories can be developed and tested against precise observations. Never before has it been possible to make such definitive observations in cosmology. Never before have cosmological discoveries arrived so fast. Never before has so much money been spent on the subject (cosmology now consumes close to a billion dollars a year). Within my living memory, our idea of the Universe was quite different from the one we have achieved today. And a few years from now, we are told, we will know to within a few percent all the important numerical parameters of our Universe. We discovered thousands of exo-planets revolving other stars. We have discovered Higgs Bozone (God Particle) in LHC (Large Hadron Collider). We have detected gravitational waves originated by black holes and neutron stars colliding. We are very close to find answers to some long pending profound questions, like - Are we alone?

Let me write down all these **Look Up questions** mapped to mid age phase.

Look Up Questions!

Question 36	Are we alone? What are we doing to detect life beyond Earth? Do we have life in our solar system?
Question 37	Do other stars also have their planetary system? How did we discover exo-planets?
Question 38	What is dark energy and dark matter?
Question 39	What is LHC and how it's important for new discoveries in astrophysics. What is God particle?
Question 40	What is LIGO and how it has detected collision of black holes and neutron stars?

Cosmic Connections which will for sure make you think!

Cosmic Connections!

Connection 26	Sharks have sixth (the Lateral Line) and seventh sense (ampullae of Lorenzini)!
Connection 27	Bio-luminosity and Bio-fluorescence life on Earth connects us to the light in the Universe!
Connection 28	Photosynthesis: How plants take light and convert it into energy? Is Quantum Biology playing a role?

"Finding theory of everything is a cosmic delusion,

which Universe keeps reminding us!"

By Kamal Deep Dham

Chapter Nine

Mature (Wisdom)

"The Universe is a symphony of strings, and the mind of God that Einstein eloquently wrote about for thirty years would be cosmic music resonating through eleven-dimensional hyper space."

By Michio Kaku

Figure: Artist's representation of parallel Universe - the Multiverse.

My Story!

Mature age relates to those who with long lives and have acquired a rich repository of experience that they can use to help guide others. You need a variety of life lessons before you can become the wise, and the respectable human being. Elders thus represent the source of wisdom that exists in each one of us, helping us to avoid the mistakes of the past while reaping the benefits of life's lessons. They become path finder for others to see the big picture.

My grandfather used to teach me science and math when I was in my primary school. He was a retired teacher of a higher secondary school in a small town of state of Haryana. Today, after years, I can say with my collected wisdom, that my interest in science and math was seeded by him. It was he who taught me Pythagoras theorem. He motivated me to do my first experiment on spectrum of light by a rainbow wheel. He taught me the concepts of Pascal's law of fluid dynamics. He taught the subject to me in such a practical way that I developed passion around it that I carry till today. That's what an old man can do to the new generation with his wisdom and experience.

Story of the Universe!

What is Cosmos? Who is governing such a huge system of energy and matter? Why it is so simple yet complex to understand? Do we all have purpose and role to play on Earth to resolve the mysteries around the Universe? Answers to these meta-physical questions can only make sense when we accumulate or seek wisdom around it.

Do we have a theory of everything? In the last few decades, string theory has emerged as the most promising candidate for a microscopic theory of everything. It is infinitely more ambitious than that: it attempts to provide a complete, unified, and consistent description of the fundamental structure of our Universe.

The essential idea behind string theory is that, all different fundamental particles of the Standard Model are just different manifestations of one basic object: a string. How can that be? Well, we would ordinarily picture an electron, for instance, as a point with no internal structure. A point cannot do anything, but move. But, if string theory is correct, then under an extremely powerful microscope we would realise that the electron is not really a point, but a tiny loop of string. A string can do something more than moving- it can oscillate in different ways. If it oscillates in a certain way, then from a distance, we are unable to tell it is really a string and assume it as an electron. But if it oscillates some other way, well, then we call it a photon, or a quark, or a ... you get the idea. So, if the string theory is correct, the entire world is made of strings!

Out here, in the mature phase, I have linked those theories that describe the bigger and unified picture of the Universe and our evolution. Most of these areas are in research state and yet to be proven experimentally, example is the String theory that is envisaged as the theory of everything. Let us think of few profound questions now. Do we live in only 3 dimensions or we have more? Do we live in one Universe of we have multiverse?

Look Up Questions!

Question 41   Do we have theory of everything (string theory) that combine theory of big (classical mechanics) and theory of small (quantum mechanics)? How many dimensions do we may have according to the string theory?

Question 42   What is Singularity?

Question 43   Is there only one Universe or we live in multiverse? What are the theories behind multiverse?

Question 44   What is time travel? Is it even possible?

Question 45   How quantum physics is changing the perception of Universe?

Cosmic Connections!

Connection 29   Quantum physics reveals a basic oneness of the Universe?

"Earlier there was only one science, no separation like

physics, chemistry and biology! We used same one science to know the

mysteries of classical reality! We have to unite sciences again today,

to know mysteries behind quantum reality!"

By Kamal Deep Dham

Chapter Ten

Death (Life)

"Death is very likely the single best invention of life. It is life's change agent. It clears out the old to make way for the new."

By Steve Jobs

"Death is a doorway that delivers you to a new life"

By Ryan Wykle

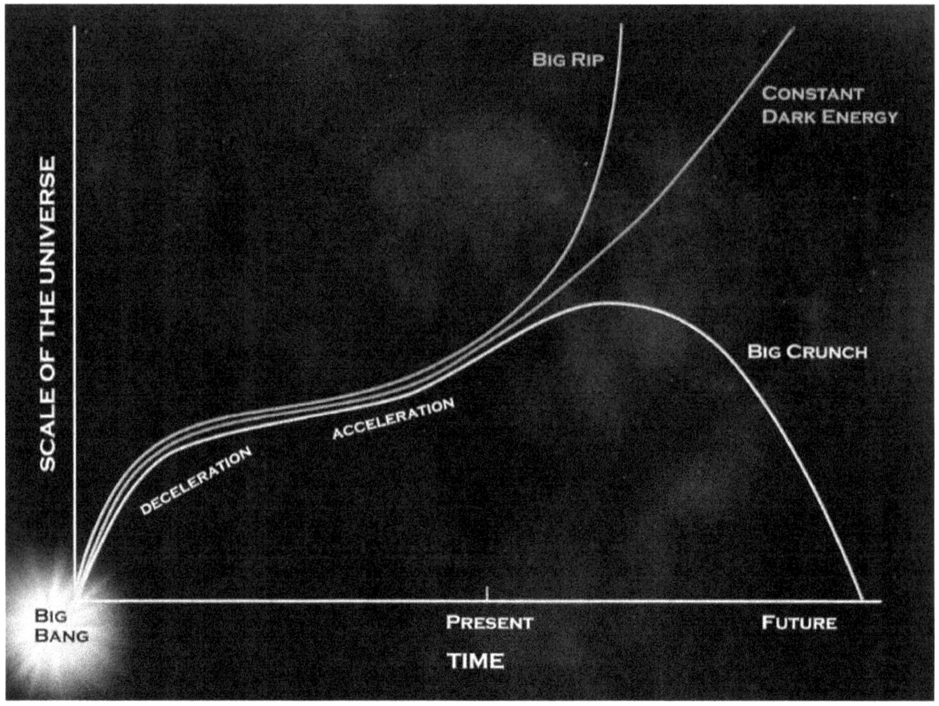

Figure: possible options for the Universe to end itself.

My Story!

Separation from someone whom you know and attached with is the most difficult emotion in one's life. Just few months back, I had to visit Delhi, to become part of death rituals of my uncle whom I was very attached to. The entire family gathered at Delhi to give tribute to the departed soul. In Hindu religion (like others) there are many defined rituals that are followed from the death time till 13th day. The key ones are:

- Day 1: Funeral of the body

- Day 2: Collection of ashes/remains from the funeral site

- Day 13: *Visarjan* (Immersion) of remains in *Ganges* (holy river)

If you become part of these rituals, you will notice that the entire process reminds you of what we are made of. The funeral of a biologically dead body converts the human cells into gasses and ashes of bones. The gases and the smoke from the burning body goes up in the sky and ashes join sand somewhere on the river shore where it is immersed. These bones and gases are further made of basic building blocks that we call atoms of oxygen, hydrogen, carbon etc.

Other religions like Islam or Christianity follow burial of the dead body in the Earth's surface. It is also an acknowledgment of our very origin and what we are made of.

Story of the Universe!

There are many theories that describe end of the Universe!

- BIG RIP

- BIG CRUNCH

- BIG FREEZE

There are no experimental observations yet that suggest the way where our Universe is marching towards.

Big Rip is a hypothetical cosmological model concerning the ultimate fate of the universe. According to this model, the matter of the universe, from stars and galaxies to atoms and subatomic particles, and even space-time itself, is progressively torn apart by the expansion of the universe at a certain time in the future.

Big Crunch theory suggests the Universe will one day stop expanding. Then, as gravity pulls in matter, the universe will begin to contract, falling inward until it has collapsed into a super-hot, super-dense singularity. The universe is like a giant soufflé. It starts out small, expands as it heats up, and eventually, as the soufflé cools, begins to collapse. Big Crunch is the consequence of the Big Bang Theory.

Another popular scenario for the end of the universe that relies on deciphering the true nature of dark energy is the Big Freeze (also referred to as Heat Death or the Big Chill).

I want to also highlight few aspects of metaphysic areas here that need further wisdom and research to get the true meaning of life and its connection with Cosmos. For example

- What is consciousness?

- Why there is always cause and effect?

- What is real meaning of death?

Like life comes with birth and then ends with death! Does Universe will also have an end or is it immortal? What is aging the Universe? How long will it survive? These are some thought provoking questions that cosmologists are still trying to find answers to. There are few un-proven theories that explain the possibilities.

Look Up Questions!

Question 46	How long the Universe will last?
Question 47	What are the different theories that explains end of the Universe?
Question 48	Is our Universe under simulation?
Question 49	What is consciousness?
Question 50	Will Universe take birth again?

Cosmic Connections!

Connection 30	Hindu religion (*Brahma*) describes cyclic creation and destruction cycle of Universe!

"Infinity is the destination of expanding Universe!"

By Kamal Deep Dham

Cosmic Connections

"We are all connected; To each other, biologically,
To the Earth chemically, To the rest of the Universe atomically.

Yes, we are part of the Universe and Universe is within us!

So, when I Look Up in the night sky, I feel big,
because the atoms in my body came from those stars!"

By Neil DeGrasse Tyson

There is a deep connection between life on the Earth and the Universe we live in. I have detailed out all the cosmic connections which you have read in the 10 chapters above. These connections are linkage between life, laws and patterns we see in nature with the Cosmos we evolved in. 90% of the connections I have chosen, have pure scientific and logical reasoning. Rest 10% is from metaphysical and spiritual space.

Cosmic Connection 1
We are connected to the Universe atomically!

We are star kids! After the Big Bang, tiny particles stitched together to form hydrogen and helium. As time went on, young stars formed when clouds of gas and dust gathered under the effect of gravity, heating up as they became denser. At the stars' cores, bathed in temperatures of over 10 million degrees C, hydrogen and then helium nuclei fused to form heavier elements. It is a reaction known as nucleo-synthesis.

Figure: We are connected to Universe atomically.

As these heavier nuclei are produced, they too are burnt inside stars to synthesise heavier and heavier elements. The elements came together and burst apart until eventually they formed all the stuff we see on the Earth today. They came together to make us, along with everyone else in the Universe.

Cosmic Connection 2
We are connected to Earth chemically!

Our human body and almost majority of life on the Earth are mainly made of Oxygen, Carbon, and Hydrogen as the key elements. The evolution of life suggests that the life started on the Earth with single cell organisms, then evolved into multi-cellular and then complex and intelligent life like us. Though we don't have clue on how the first self-replicating single cellular life started from the building blocks like Carbon, Oxygen and other molecules, but we for sure know that it was some complex chemical grand experiment way back in history that gave birth to first living cell.

Also, the human life like us, are very deeply connected to chemical outcomes of natural and artificial reactions between different elements and molecules on the Earth. The air we breathe, the food we eat, the groceries we use in our daily routine, the medicines we use to cure a disease, the car and the fuel we use to travel, the pen or pencil we use to write, the cellphone we use to communicate and so on, are all result of complex chemical reactions of elements and molecules we see on the Earth.

Cosmic Connection 3
Women's womb holds a powerful connection to the astronomical cycles of the Earth, the Sun and the Moon!

Though it's not scientifically proven, but in my view, it has a deep cosmic connection. Consider this, a women's menstrual cycles are intrinsically linked to the cycles of the Earth, Moon and Sun. The Moon cycle is 29.5 days, and the average woman's menstrual cycle is 29.5 days. Women whose cycles are closest to the 29.5 days cycle have higher rates of fertility. In addition, there are 13 Moon cycles in a calendar year, and the average age of menarche (a girl's first menstruation) is age 13. The average age of menopause is 52, which is also the number of weeks in a year. There is an average of 4 weeks to a women's menstrual cycle and 4 seasons in a year. Don't you think, there are hints that before electricity and the light bulb, shifts in the amount of light at night during the Moon's various phases could possibly have played a significant role in our menstrual cycle–maybe by affecting certain body processes impacted by light exposure, such as the production of certain hormones?

Embryonic recapitulation of all living beings on Earth, tells us how we evolved!

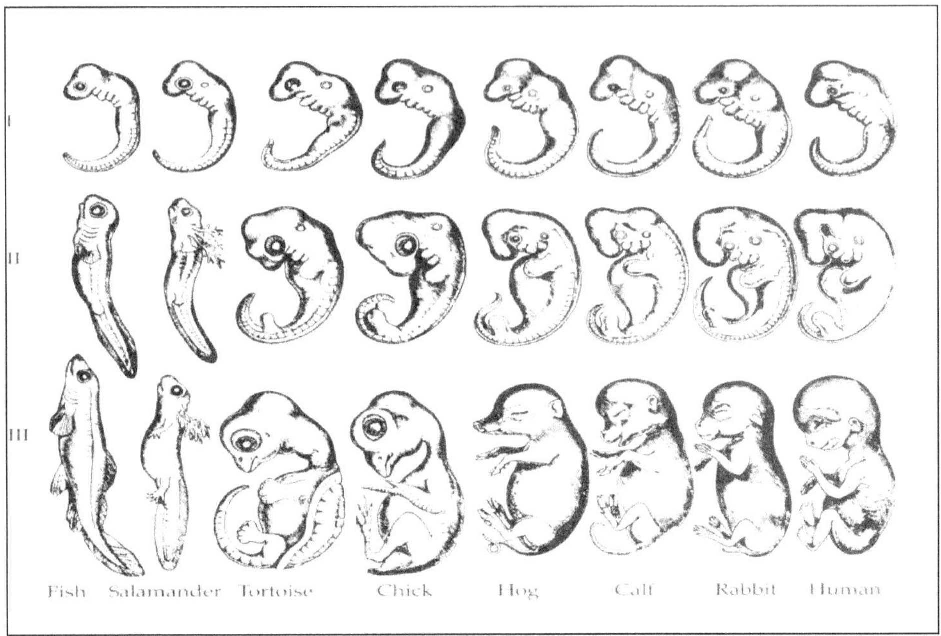

Figure: Embryonic recapitulation of Fish, Salamander, Tortoise, Chick, Hog, Calf, Rabbit, Human!

Embryonic recapitulation is the evolutionary theory that suggests that the embryos reconstruct evolutionary forms and stages during the development of the organism. This is often stated as "ontogeny recapitulates phylogeny." This theory of recapitulation confirms our connection to the Cosmos and the way we evolved from similar source and origin. Did Universe also go through recapitulation phase during its embryonic phase? We don't know! may be a good topic for astrophysicists to research!

We are connected to each other biologically!

Figure: Its 2% difference in DNA, which makes us human than chimps!

All living organisms store genetic information using the same molecules — DNA and RNA. Written in the genetic code of these molecules is compelling evidence of the shared ancestry of all living things. Evolution of higher life forms requires the development of new genes to support different body plans and types of nutrition. Even so, complex organisms retain many genes that govern core metabolic functions carried over from their primitive past. So, all the life on Earth is biologically connected. If you drill down to the DNA level, we will find that all the living beings (including plants) share common DNA.

Consider the following:

- Genome-wide variation from one human being to another can be up to 0.5% (99.5% similarity)

- Chimpanzees are 96% to 98% similar to humans, depending on how it is calculated.

- Cats have 90% of homologous genes with humans, 82% with dogs, 80% with cows, 79% with chimpanzees, 69% with rats and 67% with mice.

- Cows are 80% genetically similar to humans.

- 75% of mouse genes have equivalents in humans, 90% of the mouse genome could be lined up with a region on the human genome 99% of mouse genes turn out to have analogues in humans

- The fruit fly (Drosophila) shares about 60% of its DNA with humans (source).

- About 60% of chicken genes correspond to a similar human gene.

- About 50% of banana genes correspond to a similar human gene.

- 26% - Yeasts are single-celled organisms, but they have many housekeeping genes that are the same as the genes in humans, such as those that enable energy to be derived from the breakdown of sugars.

Don't you think we all have strong biological connections to each other?

Cosmic Connection 6
Our blood stream has heavy element like Iron!
From where did it come?

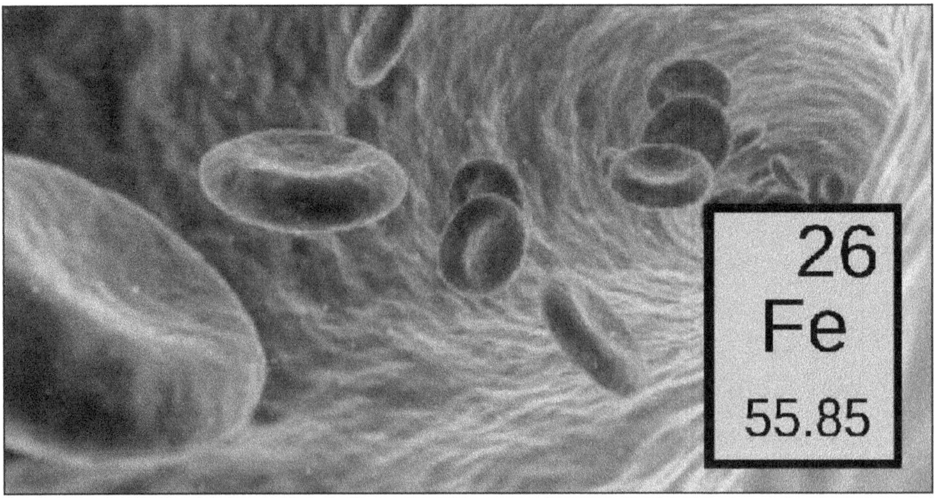

Figure: Heavy element like Iron in our blood stream makes us think, from where it came?

The Iron in our blood was formed in the stars, billions of years ago and trillions of miles away! The human body requires iron to perform many vital physiological functions. For instance, Iron is the key component of haemoglobin that allows red blood cells to transport Oxygen throughout our body, and it plays a key role in cell growth and differentiation. Low levels of Iron can leave you feeling physically tired and weak, impair mental function, and weaken the immune system. Having too much Iron in the body, on the other hand, can poison certain organs and even cause death. Maintaining the optimal balance of Iron within the body is therefore essential for one's health. This heavy element in our body connects us to the Cosmos! It's the same Iron that is formed in the stardust when old stars died.

Cosmic Connection 7
Mutation is the cause of variety of life we see on the Earth and UV light in the Universe controls it!

Mutation is a change in the sequence of an organism's DNA. What causes mutation? Mutations can be caused by high-energy sources such as radiation (UV) or by chemicals in the environment. They can also appear spontaneously during the replication of DNA. Some of these mutations are neutral (cause no significant effect on the functioning of the organism), some are deleterious (have harmful effects for the organism's body) and some are advantageous (impart the organism with some positive trait, that ensures better survival).

Another way to look at mutation is around the evolution of life on the Earth. Evolution is the gradual change in a population or species over time due to heritable changes in their DNA. Mutations are the basic process for evolution to occur. As I said mutations can be useful, deleterious or neutral. Now suppose there is a useful mutation in an organism that enables it to live longer or survive in some harsh environmental conditions, most likely this organism will be more successful reproductively and may produce more offspring. Its offspring will in turn carry this useful mutation and pass it on to their offspring. The process continues till this mutation is found in a significant proportion of the species. Over time (may take 1000s of years), this proportion increases and majority of the population or species have this mutation and is different from the earlier parent/ancestral population. This is how a population or species changes over time and over a very long period, the species may look or be very different from the parent species when it all started.

Cosmic Connection 8
Every time we take a breath, we take atoms from breath of other living beings ever lived on the Earth.

The air we breathe is composed primarily of Nitrogen and Oxygen with a small amount of other gases, including Carbon-dioxide. The individual atoms making up those molecules have been on the Earth for a long time — very little Carbon, Oxygen or Nitrogen is lost to outer space, and only the occasional meteor brings a small extraterrestrial source of these gases to our planet. So is there a chance that you are breathing in the exact same gas molecules that dinosaurs or say Julius Caesar once breathed? Answer is mostly no as these individual molecules are constantly rearranged and recycled through the biochemical and geochemical processes. So, every breath you take and every bite you swallow is composed of atoms that have been here for a long time but might not be same molecule.

For example, over hundreds of years a tree will grow, die and then decompose. As it decomposes, that atom of Carbon is released back to the atmosphere as Carbon-dioxide and is used to generate oxygen through photosynthesis.

So, every breath you take has, at one time or another, might have been associated with another living organism.

Cosmic Connection 9

Only Six degrees of separation between us to any human on Earth!

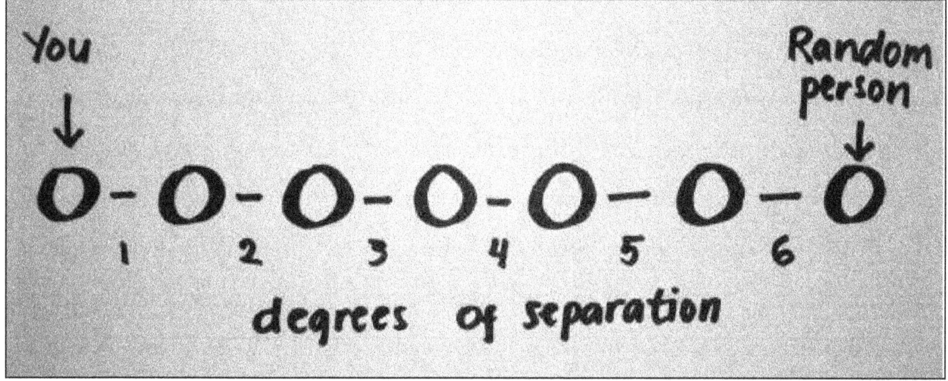

Figure: We humans are all connected by only 6 degrees of separation from others

Have you heard about the theory of "6 degrees of separation"? It suggests that any person on the planet can be connected to any other person on the planet through a chain of acquaintances that has no more than 5intermediaries. This explains that all the human beings are close cousins of each other and connected biologically.

Cosmic Connection 10

There are around 100 Trillion microbes/bacteria living in our body. These microbes were the first form of life that originated on our Earth!

It is considered that there are about 100 Trillion cells in a human body. However, one tenth of them are hardly real human cells. The human body is home of Trillions of bacteria, viruses, fungi, and other tiny organisms. These organisms are known as microbes. These microbes belonging to different communities collectively are called as microbiome. The human microbiome is a source of genetic diversity and no two human microbiomes are the same. Microbiome is an essential component of immunity and a functional entity that influences metabolism and modulates drug interactions. It has been known since long time that microorganisms in the human body play an important role in maintaining human health.

These microbes make the oldest form of life on Earth. These have been around for more than 3.5 Billion years. From the past six million years, these microbes are evolving together with humans. As these have changed over time, microbes and humans have formed complex relationships with each other. To stay healthy, humans need microbes and many microbes need specific environment provided by the human body to survive. Humans and microbes rely on these interactions to grow and stay healthy. Isn't it fascinating to know that one microbe Billions of years ago started life on Earth and now each one of us has Trillions of them in our body!

Cosmic Connection 11
Human brain mirrors visual phenomenon in the natural Universe!

The idea of the Universe as a 'giant brain' has been proposed by scientists and science fiction writers for decades. But now physicists say there may be some evidence that it's true in a sense.

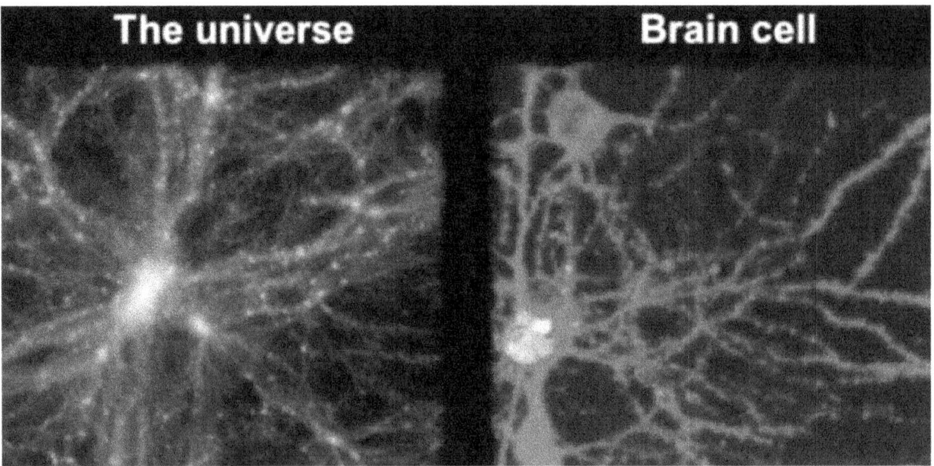

Figure: Similarity between Universe and Brian cell

As per a study published in Nature's Scientific Reports, the Universe may be growing in the same way as a giant brain – with the electrical firing between brain cells 'mirrored' by the shape of expanding galaxies. The results of a computer simulation suggest that "natural growth dynamics" – the way that systems evolve are the same for different kinds of networks whether it's the internet, the human brain or the Universe as a whole.

Cosmic Connection 12
Mathematics is the languages of Universe!

The idea that everything is, in some sense, Mathematical goes back at least to the Pythagoreans of ancient Greece and has spawned centuries of discussion among physicists and philosophers. In the 17th century, Galileo famously stated that our Universe is a "grand book" written in the language of Mathematics. When you look around, do you see any geometric patterns or shapes? Try throwing a pebble, and watch the beautiful shape that nature makes for its trajectory!

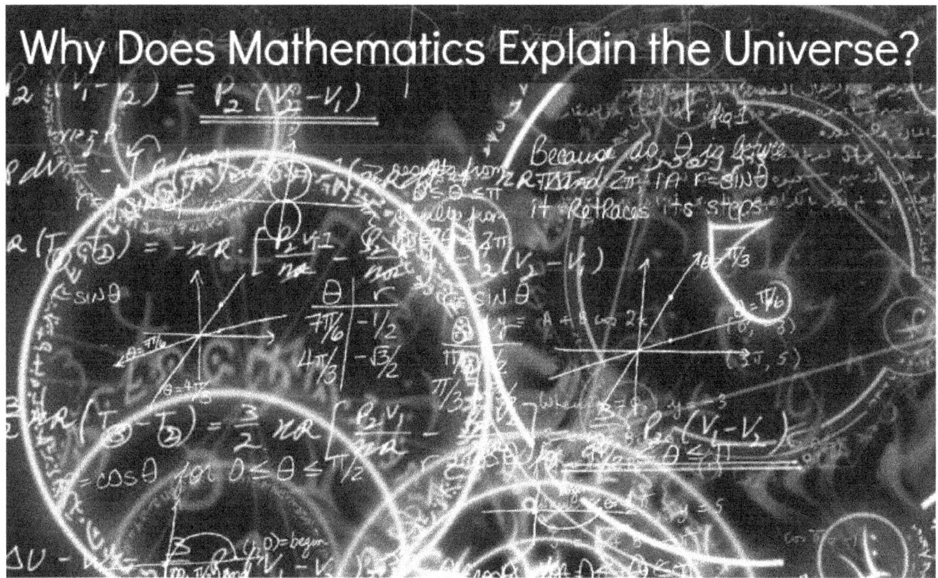

Figure: Universe can be explained in the language of mathematics.

The trajectories of anything you throw have the same shape, called an upside-down parabola. When we observe how things move around in orbits in space, we discover another recurring shape: the ellipse. Moreover, these two shapes are related: So, in fact, all trajectories are simply parts of ellipses.

We humans have gradually discovered many additional recurring shapes and patterns in the nature, involving not only motion and gravity, but also electricity, magnetism, light, heat, chemistry, radioactivity and subatomic particles. These patterns are summarized by what we call our laws of Physics. Just like the shape of an ellipse, all these laws can be described using mathematical equations. Equations aren't the only hints of mathematics that are built into nature, there are also numbers.

There's something very mathematical about our Universe, and the more carefully we look, the more math we seem to find. So, what do we make of all these hints of mathematics in our physical world? Look at the next connection to see this strong bond between Mathematics and the universe!

Cosmic Connection 13
Fibonacci Series and Golden Ratio is everywhere in the Universe!

$$F_n = F_{n-1} + F_{n-2}$$

The **Fibonacci sequence** is a set of numbers that starts with a one or a zero, followed by a one, and proceeds based on the rule that each number (called a Fibonacci number) is equal to the sum of the preceding two numbers. If the Fibonacci sequence is denoted as F (n), where n is the first term in the sequence, the following equation obtains for n = 0, where the first two terms are defined as 0 and 1 by convention:

F (0) = 0, 1, 1, 2, 3, 5, 8, 13, 21, 34 ...

A ratio, related to Fibonacci series is the Golden ration. Let us take a look at this fascinating ratio.

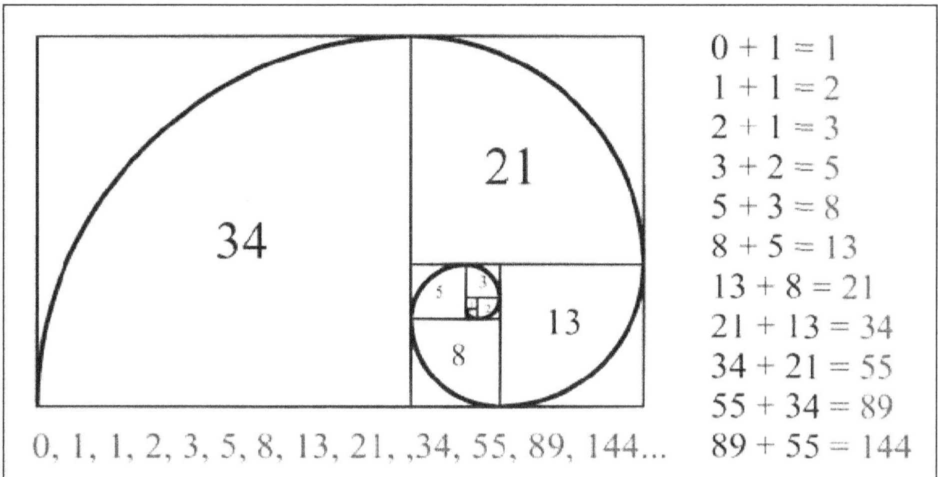

Figure: Fibonacci sequence represented geometrically.
Spiral arc is the outcome as we can see in the picture.

The **Golden ratio** is a special number found by dividing a line into two parts so that the longer part divided by the smaller part is also equal to the whole length divided by the longer part. It is often symbolized using phi, after the 21st letter of the Greek alphabet. In an equation form, it looks like this:

a/b = (a+b)/a = 1.61803398874989484820 …

As with pi (the ratio of the circumference of a circle to its diameter), the digits go on and on, theoretically into infinity. Phi is usually rounded off to 1.618. This number has been discovered and rediscovered many times, which is why it has so many names — the Golden mean, the Golden section, divine proportion, etc. Historically, the number can be seen in the architecture of many ancient creations, like the Great Pyramids and the Parthenon. Fibonacci sequence and golden ratio is there everywhere. Let's list down some interesting examples that we see and encounter in our life and often miss to establish the linkage!

a) Flower petals

Figure: Fibonacci sequence and Phi in flower petals.

The number of petals in a flower consistently follows the Fibonacci sequence. Famous examples include the Lily thathas three petals, Buttercups that have five, the Chicory's 21, the Daisy's 34, and so on. Phi appears in petals because the ideal packing arrangement as selected by Darwinian processes, each petal is placed at 0.618034 per turn (out of a 360° circle) allowing for the best possible exposure to Sunlight and other factors.

b) Flower Pistils

Figure: Fibonacci sequence and Phi in flower pistils.

The head of a flower is also subject to Fibonaccian processes. Typically, seeds are produced at the centre, and then migrate towards the outside to fill all the space. Sunflower provides a great example of these spiralling patterns.

In some cases, the seed heads are so tightly packed that total number can get quite high, as many as 144 or more. And when counting these spirals, the total tends to match a Fibonacci number. Interestingly, a highly irrational number is required to optimize the filling. Phi fits the bill rather nicely.

c)   Pinecones

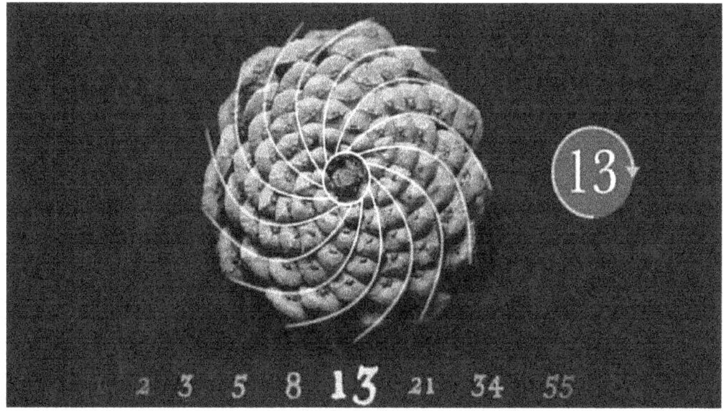

Figure: Fibonacci sequence in pinecones.

Similarly, the seed pods on a pinecone are arranged in a spiral pattern. Find a pine cone and look at it from the bottom. Count the number of spirals going from the centre of the cone (where it attached to the tree) to the outside edge. Count the spirals in both directions. The resulting numbers are usually two consecutive Fibonacci numbers (e.g., 0, 1, 1, 2, 3, 5, 8, 13, 21, 34, . . .)

d) Tree Branches

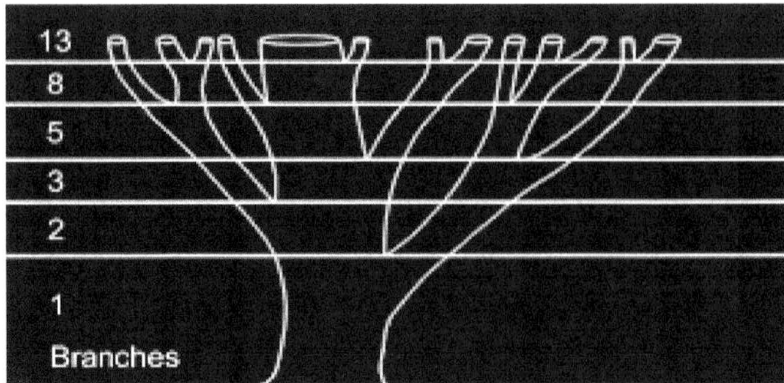

Figure: Fibonacci sequence in tree branches arrangement.

The Fibonacci sequence can also be seen in the way tree branches form or split. A main trunk will grow until it produces a branch, which creates two growth points. Then, one of the new stems branches into two, while the other one lies dormant. This pattern of branching is repeated for each of the new stem. A good example is the Sneezewort. Root systems and even algae exhibit this pattern.

e) **Leaves**

Figure: Fibonacci sequence in leaves

Leaves follow Fibonacci sequence when growing off branches and stems and in their vein patterns. I personally find the veins much more interesting and amazing to look at. Like a tree, leaf veins branch off more and more in the outward proportional increments of the Fibonacci sequence.

f) Shells

Figure: Golden ratio in shells

Snail shells and nautilus shells follow the logarithmic spiral (designed on golden ratio), as does the cochlea of the inner ear. It can also be seen in the horns of certain goats, and the shape of certain spider's webs.

g) Spiral Galaxies

Not surprisingly, spiral galaxies also follow the familiar Fibonacci pattern. The Milky Way has several spiral arms, each of them a logarithmic spiral of about 12 degrees. As an interesting aside, spiral galaxies appear to defy Newtonian physics.

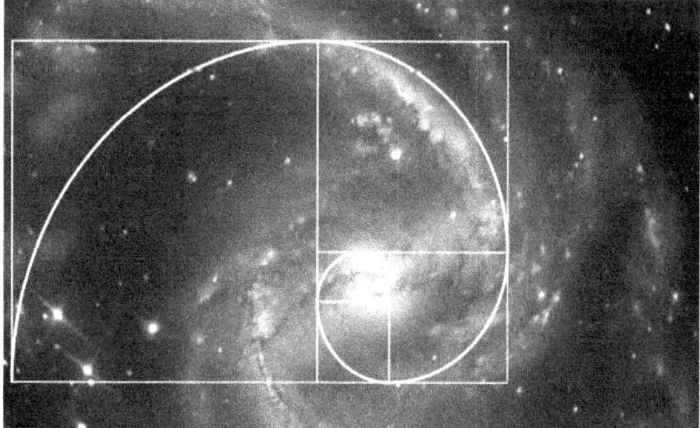

Figure: Fibonacci pattern and golden ration in spiral galaxies.

As early as 1925, astronomers realized that, since the angular speed of rotation of the galactic disk varies with distance from the centre, the radial arms should become curved as galaxies rotate. Subsequently, after a few rotations, spiral arms should start to wind around a galaxy. But they don't — hence the so-called winding problem. The stars on the outside, it would seem, move at a velocity higher than expected — a unique trait of the Cosmos that helps preserve its shape.

h) Hurricanes

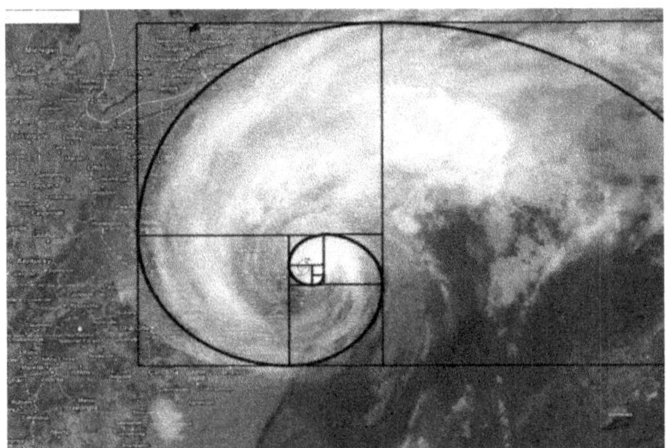

Figure: Golden ration in hurricane

We can see golden ratio in Earth's weather too. For example, much like shells and spiral galaxies, hurricanes often displays the Golden spiral.

i) Faces

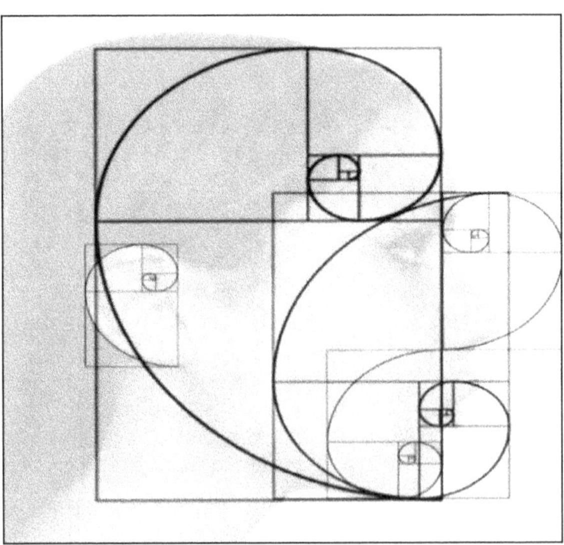

Figure: Golden ration in human face.

Faces, both human and non-human, abound with examples of the golden ratio. The mouth and nose are each positioned at golden sections of the distance between the eyes and the bottom of the chin. Similar proportions can be seen from the side, and even the eye and ear itself (which follows along a spiral).

It's worth noting that every person's body is different, but that averages across populations tend towards phi. It has also been said that the more closely our proportions adhere to phi, the more "attractive" those traits are perceived. As an example, the most "beautiful" smiles are those in which central incisors are 1.618 wider than the lateral incisors, which are 1.618 wider than canines, and so on. It's quite possible, that we are primed to like physical forms that adhere to the golden ratio — a potential indicator of reproductive fitness and health.

j) Reproductive dynamics

Speaking of honey bees, they follow Fibonacci in other interesting ways. The most profound example is by dividing the number of females in a colony by the number of males (females always outnumber males). The answer is typically something very close to 1.618. In addition, the family tree of honey bees also follows the familiar pattern. Males have one parent (a female), whereas females have two (a female and male). Thus, when it comes to the family tree, males have 2, 3, 5, and 8 grandparents, great-grandparents, gr-gr-grandparents, and gr-gr-gr-grandparents respectively. Following the same pattern, females have 2, 3, 5, 8, 13, and so on. And as noted, bee physiology also follows along the Golden Curve rather nicely.

k) DNA Molecules

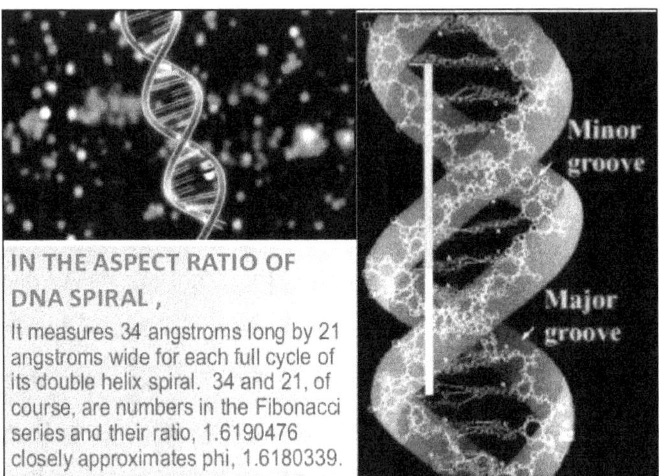

Figure: Golden ration in DNA molecules.

Even the microscopic realm is not immune to Fibonacci. The DNA molecule measures 34 angstroms long by 21 angstroms wide for each full cycle of its double helix spiral. These numbers, 34 and 21, are numbers in the Fibonacci series, and their ratio 1.6190476 closely approximates Phi, 1.6180339.

Cosmic Connection 14
Pi is not just a collection of random digits. Pi is a journey!

Pi — written as the Greek letter for p, or π — is the ratio of the circumference of any circle to the diameter of that circle. Regardless of the circle's size, this ratio will always equal pi. In decimal form, the value of pi is approximately 3.14. But pi is an irrational number, meaning that its decimal form neither ends (like 1/4 = 0.25) nor becomes repetitive (like 1/6 = 0.166666...). (To only 18 decimal places, pi is 3.141592653589793238.) Hence, it is useful to have shorthand for this ratio of circumference to diameter. According to Petr Beckmann's A History of Pi, the Greek letter π was first used for this purpose by William Jones in 1706, probably as an abbreviation of periphery, and became standard mathematical notation roughly 30 years later.

Figure: Pi is everywhere we see in nature around us.

Whether or not humans and gods grasp the irrational number, pi seems to crop up everywhere, even in places that have no connection to circles. For example, among a collection of random whole numbers, the probability that any two numbers have no common factor — that they are "relatively prime" — is equal to $6/\pi^2$. Strange, no?

But pi's ubiquity goes beyond math. The number crops up in the natural world, too. It appears everywhere there's a circle, of course, such as the disk of the Sun, the spiral of the DNA double helix, the pupil of the eye, the concentric rings that travel outward from splashes in ponds. Pi also appears in the physics that describes waves, such as ripples of light and sound. It even enters into the equation that defines how precisely we can know the state of the Universe, known as Heisenberg's uncertainty principle.

Finally, pi emerges in the shapes of rivers. A river's windiness is determined by its "meandering ratio" or the ratio of the river's actual length to the distance from its source to its mouth as the crow flies. Rivers that flow straight from source to mouth have small meandering ratios, while ones that lollygag along the way have high ones. Turns out, the average meandering ratio of rivers approaches — you guessed it — pi.

Albert Einstein was the first to explain this fascinating fact. He used fluid dynamics and chaos theory to show that rivers tend to bend into loops. The slightest curve in a river will generate faster currents on the outer side of the curve, which will cause erosion and a sharper bend. This process will gradually tighten the loop, until chaos causes the river to suddenly double back on itself, at which point it will begin forming a loop in the other direction.

Cosmic Connection 15

Prime Numbers! It was all so neat and elegant. Numbers that refuse to cooperate that don't change or divide, numbers that remain themselves for all eternity.

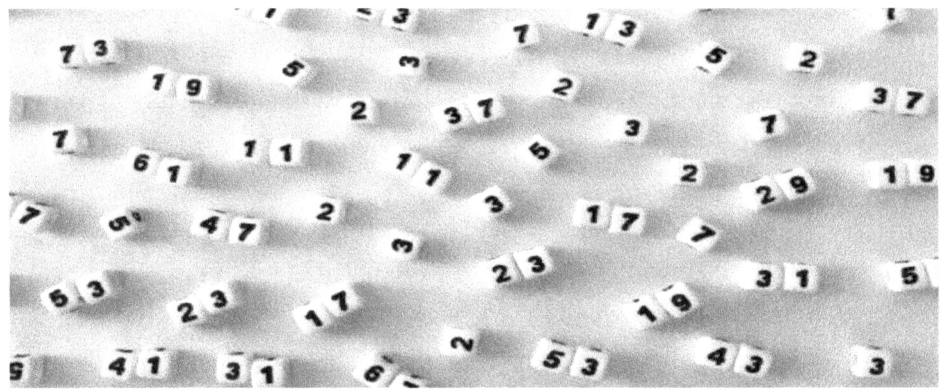

Figure: Prime numbers are everywhere around us!

A prime number is an integer, or whole number, that has only two factors - 1 and itself. Put another way, a prime number can be divided evenly only by 1 and by itself. Prime numbers also must be greater than 1. For example, 3 is a prime number, because 3 cannot be divided evenly by any number except for 1 and 3. However, 6 is not a prime number, because it can be divided evenly by 2 or 3.

As we study nature, we discover more and more examples where prime numbers are visible! Let's see some examples:

- It is well-known that a certain kind of locust (a cicada) spends most of its time in hiding, only reappearing to mate every 13 or 17 years (13 years in the southern U.S., 17 years in the northern US). There are about 1,500 species of cicadas known. There are those that appear yearly in midsummer, and there are also the so-called "periodic" cicadas. They appear at prime number intervals - 7 years, 13 years and 17 years.

- Prime numbers in living beings: Number of chromosomes in human is 23 and number of chromosomes in a Sunflower is 17, both are prime numbers. Most flowers have an odd and often prime number of petals: five is a common number. Sunflower has 23 petals. Starfish has 5 points.

Cosmic Connection 16
Power of Symmetry: Nature and Universe both loves symmetry!

Figure: Symmetry everywhere in the Universe!

An object exhibits symmetry if it looks the same after a transformation, such as reflection or rotation. Symmetry is the underlying mathematical principle behind all patterns and is important in art (used in architecture, pottery, quilting and rug making). It is observed in mathematics (relating to geometry, group theory and linear algebra), biology (in shapes of organisms), chemistry (in shapes of molecules and crystal structures), and physics (where symmetries correspond to conserved quantities).Everywhere we look, our eyes are drawn first to the patterns of symmetry that exist, and that the object itself is a secondary consideration. Let's see some examples in shapes of plants, birds and animals:

- Plant leaves pattern have symmetry inbuilt.
- Flowers offer a variety of radial symmetry.
- Stripes on tiger and zebra faces make us think why they are so symmetric.
- Bees form their honeycombs in a pattern as well. There seems to be a lot of hexagonal symmetry in nature.
- We have already seen symmetrical shapes of galaxies in the Universe.

Cosmic Connection 17

Periodic Table symmetry: The element of surprise was not allowed near the Periodic Table!

Figure: Symmetry in periodic table.

The periodic table of elements arranges all the known chemical elements in an informative array. Elements are arranged from left to right and top to bottom in order of increasing atomic number. Order generally coincides with increasing atomic mass.

It is the symmetry in the periodic table that connects us to the Universe. There are special relation and commonality between the elements grouped under different segments (as in figure with different colours). This symmetry helped in the prediction of new elements and its properties before it was discovered.

The rows are called periods. The period number of an element signifies the highest energy level an electron in that element occupies. The number of electrons in a period increases as one move down the periodic table; therefore, as the energy level of the atom increases, the number of energy sub-levels per energy level increases.

Elements that occupy the same column on the periodic table (called a "group") have identical valance electron configurations and consequently behave in a similar fashion chemically. For instance, all the group 18 elements are inert gases.

Cosmic Connection 18
Everything Spins - around the Nucleus and in an orbit!

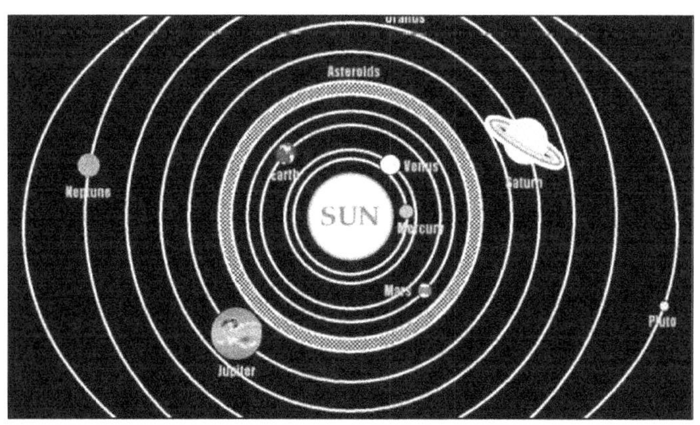

Figure: Large bodies spin around common centre.

Everything in the Cosmos is constantly spinning. Nucleus, Electrons, Planets, Stars, Galaxies everything appears to be in an everlasting motion. Some spin at lower velocities and other sat velocities close to the speed of light. But why is everything spinning?

Let's begin four and half billion years ago, when our Solar System began to form around clouds of Helium and Hydrogen. As gases of varying densities moved around the space, a triggering event (maybe a supernova) caused these gases to coalesce. As the gravity of these bodies increased, they began pulling in everything around them, causing these bodies to spin. A curious phenomenon here is that, whenever this happens, the spin appears to move in the same direction: Counter-clockwise.

Figure: Small elementary particles spin around nucleus and its axis

Eventually, all these elements collide and form larger and larger bodies that become planets, Moons, etc., all the while maintaining the angular momentum as inertia keeps the spin going. This is what causes the spin, and it's the reason why everything appears to follow the same kind of spin. Is it slowing down? Yes, but at a slower rate than humans can consciously detect.

Even when we go to atomic level, we see there too elementary particles spinning around the nucleus. Electrons spin around the nucleus. Every elementary particle has its own spin on its axis.

Cosmic Connection 19
Cycles everywhere: Universe will keep hitting you with same lessons, until you learn!

Where ever you see, you will find cycles in the Universe we live in. Consider the following:

- Elementary particles cycles around its nucleus.
- Moon cycles around the Earth in 29.5 days.
- Earth cycles around its axis in 24 hours and around the Sun in 365 days.
- Being part of the Milky Way galaxy, sun being at its exterior arm revolves around the supermassive black hole at the galaxy's centre.
- All the 8 planets in the solar system orbit around the Sun.
- Seasons cycle on the Earth.
- Day and night cycle on the Earth.
- Life on Earth has many cycles built in. Example birth and then death, work and sleep, menstrual cycle in females, hibernation cycle in many animals on Earth, migration of birds is cyclic too, many plants, fruits and crops are seasonal, etc.

The more you explore the more you explore cyclic nature of the Universe and its connection to life we see on our Earth.

Cosmic Connection 20
Water has memory: All water has perfect memory and is forever trying to get back to where it was!

Though it's not proven scientifically, but we do have thought provoking observations around this fact. Water memory is the ability of water to retain a memory of substances previously dissolved in it even after an arbitrary number of serial dilutions. It has been claimed to be a mechanism by which homeopathic remedies work, even though they are diluted to the point that no single molecule of the original substance remains.

Water memory defies conventional scientific understanding of physical chemistry knowledge and is not accepted by the scientific community.

Cosmic Connection 21
Moon creates tides in the ocean and
Sun controls the seasons on the Earth!

Tides are a phenomenon on Earth that occurs in a pattern that can be predicted. Tides are the rising and falling of the sea levels every day. Most places near the ocean experience two high tides and two low tides every 24 hours.

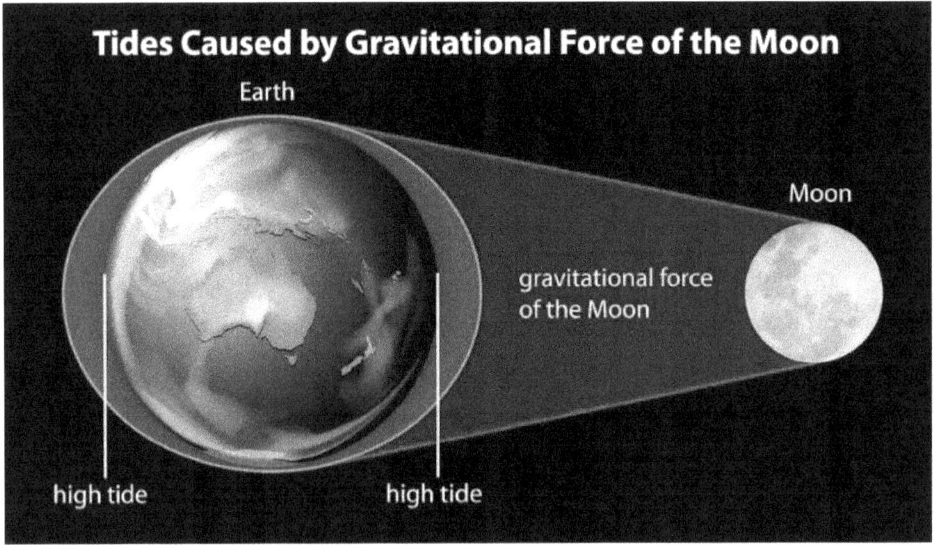

Figure: Moon's control on earth.

Tides are caused by the pull of gravity – mostly from the Moon. The Moon's gravity pulls water away from the Earth's surface. This causes the water to rise, forming a bulge of water in the oceans. On the opposite side of Earth (away from the Moon), the water is also pulled away from Earth's surface forming another bulge. These bulges form high tides. The part of Earth closest to the Moon usually has the highest tides.

Low tides occur where the water has been pulled away. The Sun's gravity also influences the tides, but the Moon has a greater effect than the Sun because the Moon is so much closer to the Earth.

The seasons occur because Earth's axis is tilted. The part of the Earth tilted towards the sun has summer. This half of the Earth receives more direct sunlight, which causes it to be warmer. The days are longer during summer. In the image, it is summer in the northern half of Earth. It is winter in the southern half of Earth. The sun's rays are not very direct.

Cosmic Connection 22
Why Plants/flowers face towards Sunlight?

Plants kept inside a room always grow in the direction of the window. In woodlands where there is a thick canopy of trees and Sunlight rarely falls on the Earth, very few plants survive, whether or not they require sunlight to make their food.

People have long wondered about this phenomenon until the answer was discovered and explained by the English naturalist Charles Darwin. He demonstrated that the growing shoot of a grass seedling always bends towards light. This is due to a phenomenon called phototropism.

We know that green plants make their food by the process of photosynthesis. The leaves of the green plants contain a green pigment called chlorophyll. Chlorophyll converts water from the soil and carbon dioxide from the air into oxygen and sugar.

Oxygen is then released into the air and sugar is taken as food by the plant. But this entire process of making food can only be done in the presence of

Sunlight. The leaves, therefore, lean towards the light for photosynthesis. Without Sunlight green plants cannot survive.

Figure: Plants/flowers face towards Sunlight.

How does a rigid plant bend or grow to one side? Plant cells contain a substance called Aurins. This substance tends to move away from the light. Aurins make the cells on the darker side grow faster than the cells on the side having light.

This causes the stems and the leaves to bend or lean towards the side having light. The concentration of Aurins on the dark side may be because Sunlight slows or kills these Aurins when light falls on them. However, if one part of the plant gets Sunlight, it can make food and the whole plant survives.

Cosmic Connection 23
Biological Clock is tuned to 24 hours on Earth!

Biological clock or circadian rhythm is a roughly 24-hour cycle in the physiological processes of living beings on Earth, including plants, animals, fungi and cyanobacteria. In a strict sense, circadian rhythms are endogenously generated, although they can be modulated by external cues such as sunlight and temperature. Circadian rhythms are important in determining the sleeping and feeding patterns of all animals, including human beings. There are clear patterns of brain wave activity, hormone production, cell regeneration and other biological activities linked to this daily cycle. Don't you think there is strong connection between the Earth rotation to its axis and the biological clock we see in life on the Earth?

Cosmic Connection 24
How birds can navigate long distances?
Does Earth magnetic field help them?

About 50 animal species, ranging from birds and mammals to reptiles and insects, use Earth's magnetic field for navigation. Earth's magnetic field is very weak. It ranges from approximately 30 to 60 millionths of one Tesla. By comparison, magnetic resonance imaging, or MRI, uses magnetic fields from 1.5 to 3.0 tesla.

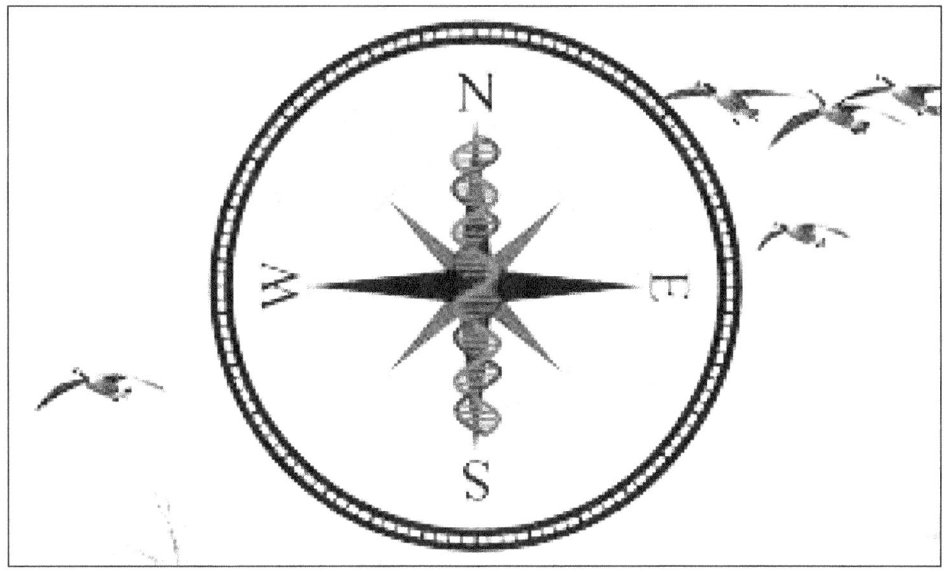

Figure: Earth gravity felid may be helping birds to navigate long distances.

New research finds that a photochemical compass may simulate how migrating birds use the magnetic field along with light, to navigate.

One theory for how it all works has been that photoreceptors in a bird's retina absorb light that causes a chemical reaction that in turn produces a short-lived photochemical species whose lifetime is sensitive to the magnitude and direction of a weak magnetic field. The idea is supported by the fact that blue light photoreceptors have been detected in retinas of migratory birds when they perform magnetic orientation. However, it has not been confirmed that a magnetic field as weak as Earth's can produce detectable changes within a photochemical molecule; nor has a photochemical molecule been shown to respond to the direction of such a magnetic field.

Cosmic Connection 25
All skin colours, whether light or dark are not due to race but due to the adaptation of life under the Sun!

Figure: Skin colours, whether light or dark, are not due to race but to adaptation of life under the Sun!

Why do people from different parts of the world have different coloured skin? Why do people from the tropics generally have darker skin colour than those who live in colder climates? Variations in human skin colour are adaptive traits that correlate closely with geography and the Sun's ultraviolet (UV) radiation.

As early humans moved into hot, open environments in search of food and water, one big challenge was keeping cool. The adaptation that was favoured involved an increase in the number of sweat glands on the skin while at the same time reducing the amount of body hair. With less hair, perspiration could evaporate more easily and cool the body more efficiently. But this less-hairy skin was a problem because it was exposed to a very strong Sun, especially in lands near the equator. Since the strong Sun exposure damages the body, the solution was to evolve skin that was permanently dark so as to protect against the Sun's more damaging rays.

Melanin, the skin's brown pigment, is a natural Sunscreen that protects tropical peoples from the many harmful effects of ultraviolet (UV) rays. UV rays can, for example, strip away folic acid, a nutrient essential to the development of healthy foetuses. Yet when a certain amount of UV rays penetrates the skin, it helps the human body use vitamin D to absorb the calcium necessary for strong bones. This delicate balancing act explains why the people that migrated to colder geographic zones with less Sunlight developed lighter skin colour. As people moved to areas farther from the equator with lower UV levels, natural selection favoured lighter skin which allowed UV rays to penetrate and produce essential vitamin D. The darker skin of people who lived closer to the equator was important in preventing folate deficiency. Measures of skin reflectance, a way to quantify skin colour by measuring the amount of light it reflects, in people around the world support this idea. While UV rays can cause skin cancer, because skin cancer usually affects people after they have had children, it likely had little effect on the evolution of skin colour because evolution favours changes that improve reproductive success.

Cosmic Connection 26
Sharks have a sixth (the Lateral Line) and a seventh sense (ampullae of Lorenzini)!

Sharks have amazing extra senses that are based on the electromagnetic field and the waves in water.

Sharks have a sixth sense called lateral line system along their sides that detects water movements. This helps the shark find prey and navigate around other objects at night or when water visibility is poor. The lateral line system is made up of a network of fluid-filled canals beneath the shark's skin. Pressure waves in the ocean water around the shark vibrate this liquid. This in turn is transmitted to jelly in the system, which transmits to the shark's nerve endings and the message is relayed to the brain.

What is the seventh sense then? Sharks can also pick up on the small electrical fields generated by other animals. Near the nostrils, around the head and on the underside of the snout, or rostrum, are small pores called ampullae of Lorenzini. Connected to these pores are long, jelly-filled bulbs that lead to nerves below the skin. Electrical signals coming from muscle movements of other organisms are received by the ampullae and transmitted through jelly-filled bulbs where they strike the nerves and signal the brain. When light is scarce in murky water or at depths, and vision is impaired, this electromagnetic sense is especially useful for locating prey. Among skates, close relatives of rays, those that inhabit deeper habitats have larger ampullae than shallow-water skates (Raschi 1986). This difference in ampulla size is probably compensated by deep-water skates for their reduced ability to depend on vision for hunting. Not only is electroreception used to locate other sharks or prey, but it is also employed as a compass during migration.

Cosmic Connection 27
Bioluminescence and Bio-fluorescence life on Earth connects us to light in the Universe!

Bioluminescence and Bio-fluorescence are two different naturally occurring phenomena seen within certain organisms and generally involve the production or emission of light. Even though they are very similar, bioluminescence and bio-fluorescence are created differently and have several factors that make them distinct. These two phenomena for sure connect life on Earth to the light we have in the Universe.

Bioluminescence is a chemical reaction created in the bodies of certain organisms and it is similar to the reaction that occurs when you crack a glow stick. This reaction doesn't produce a lot of heat so it is sometimes called "cold light".

Figure: Example of bio luminous creature in deep sea.

The uses of this phenomenon are as varied as the types of creatures capable of producing it. Fireflies use their abilities to luminescent—to blink in specific patterns—to communicate with each other. Deep sea fish like the Angler fish use their bioluminescence to attract or mimic prey, and locate food. Other animals release bioluminescent fluid as self-defence.

Bio-fluorescence on the other hand is not a chemical reaction, it is a phenomenon by which an organism absorbs blue light and emits it as a different colour, usually red, orange or green. Until recently it was thought that only certain marine organisms like corals and jellyfish had the capability to fluoresce.

Figure: Bio-fluorescence seen in life on Earth

Now, marine biologists have discovered that a wide distribution of fish also have this ability. It's thought that this unusual occurrence is used primarily for communication, camouflage and mating purposes but new research suggests many additional functions and there is much more to discover about this phenomenon. These new discoveries may one day be used to develop new products or even be used in the field of medicine to cure diseases!

Cosmic Connection 28
Photosynthesis: How plants take light and convert it into energy? Is Quantum Biology playing a role?

We all probably learnt about photosynthesis, how plants turn Sunlight into energy, in school. It might seem, therefore, that we figured out this bit of the world. But scientists are still learning new things about even the most basic stuff and photosynthesis is no different.

In particular, per a study published by an international team of scientists showed that molecules involved in photosynthesis display quantum mechanical behaviour. Not only will it help us better understand plants, Sunlight and everything in between, but it could also mean cool new tech in the future.

With photosynthesis, scientists show for the first time that there are quantum effects in living systems. This could lead to better solar panels, energy storage or even quantum computers.

I am not capturing details of how actually plants do this under Quantum rules. If you are interested, there is enough study material available on the internet.

Cosmic Connection 29
Does quantum physics revel a basic oneness of the Universe?

It results in what may appear to be some very strange conclusions about the physical world. At the scale of atoms and electrons many of the equations of classical mechanics that describe how things move at everyday sizes and speeds, cease to be useful. In classical mechanics, objects exist in a specific place at a specific time. However, in quantum mechanics, objects instead exist in a haze of probability; they have a certain chance of being at point A, another chance of being at point B and so on.

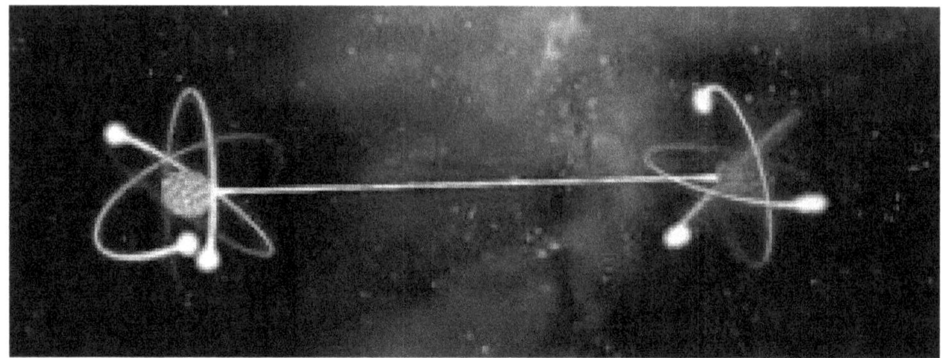

Figure: Quantum entanglement

Classical mechanics is the study of the motion of large bodies in our Universe and Quantum mechanics is the branch of physics relating to the very small.

Quantum entanglement is one of the key phenomena suggested in theory of Quantum mechanics. Quantum entanglement is a physical phenomenon that occurs when pairs or groups of particles are generated, interact, or share spatial proximity in ways such that the quantum state of each particle cannot be described independently of the state of the other(s), even when the particles are separated by a large distance — instead, a quantum state must be described for the system as a whole. There is much more to the Quantum world, but point that I want to bring out here is that how in future it may have connections to realty around us. There are studies that suggest (not yet proven though) that brain works with quantum principles.

Consciousness of living beings has connection with quantum physics. There are studies that propose the entangled brains. We have seen already possible role of quantum principles in Photosynthesis. What role quantum laws play in nature is long journey for scientist to explore, but we are already getting hints that there are connections that may explain the big questions.

Cosmic Connection 30
Hindu religion *(Brahma)* describes cyclic creation and destruction cycle of Universe!

The Hindu tradition perceives the existence of the cyclical nature of the Universe and everything in it. The Cosmos follows one cycle within a framework of cycles. It may have been created and reach an end, but it represents only one turn in the perpetual "wheel of time", that revolves infinitely through successive cycles of creation and destruction. This is the closest belief we have in cosmology with the Big Bang theory. Does this mean that we may have some connection in mythology to search for the answers?

Lookup Questions

*As promised in this section, I will answer all the 50 **Look Up** questions that we have seen in the 10 phases of human and Universe life.*

I will try to give my best shot to explain in such a way that anyone, (who has little knowledge of science) can easily comprehend the explanation.

Idea is not to get to detail, rather give the bigger picture and key pointers.

I will leave it to your Google skills to find out details if you want to further deep dive.

Let's start!

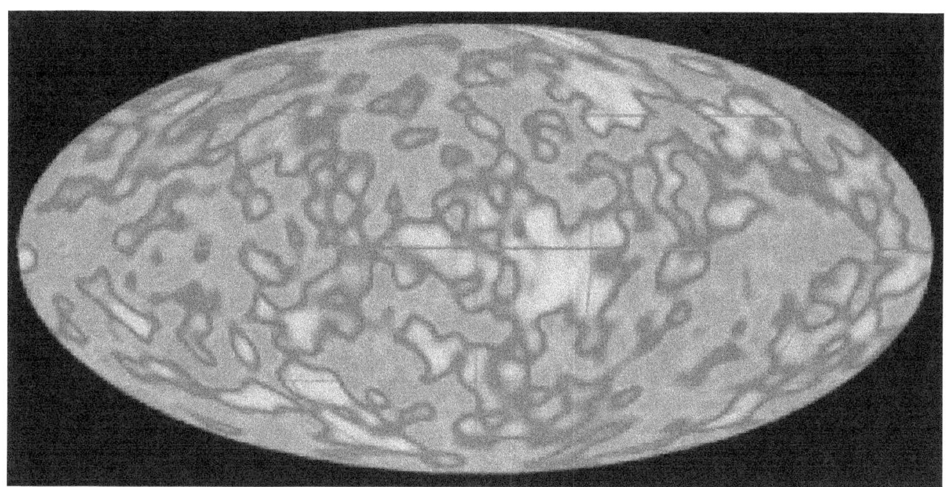

Figure: Baby universe as seen in CMB.

The Universe is everything we can touch, feel, sense, measure or detect. It includes living things like us, animals, plants, non-living things like water, air, rocks, planets, stars, galaxies, dust clouds, light, energy and even time.

The Universe contains billions of galaxies, each containing billions of stars. The Universe is incredibly huge. No one knows the exact size of the Universe, because we cannot see the edge – if there is one. All we do know is that the visible Universe is at least 93 billion light years across.

As per physics and our current understanding, the Universe consists of space time, forms of energy (including electromagnetic radiation and matter), and the physical laws that relate them. The Universe encompasses all of life, all of history, and some philosophers and scientists suggest that it even encompasses ideas such as mathematics and logic.

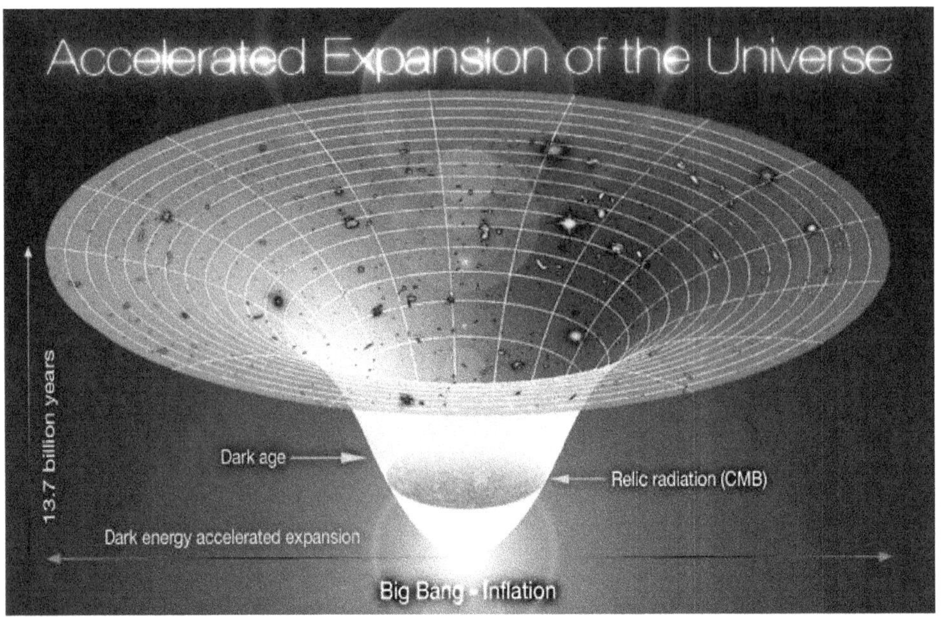

Figure: Expanding universe as seen by Hubble.

Before the 1920s, beginning of the Universe had no relevance in scientific understanding. Astronomers believed the Cosmos to be eternal and unchanging. We knew of only one galaxy and a few million visible stars, and this was the scope of our observable Universe. Then astronomer Edwin Hubble observed distant galaxies speeding away from each other and formulated Hubble's Law to explain the Universe's uniform expansion. Based on Hubble's observation, scientists proposed expanding model of the Universe as part of the Big Bang theory. The Big Bang theory is most widely accepted theory that explains birth of the Universe. Through its simplest explanation, that says the Universe as we know started with a small singularity, then expanded over the next 13.7 billion years to the Cosmos that we live in today.

Look Up Question 2
What was there before the beginning (The Big Bang)?

We don't have a clear picture yet about what was there before the Big Bang. There are many unproven theories that suggest possible answers, but we must wait to get the final one.

As of now, scientists believe that before the Big Bang there was nothing, no such thing as time or space. There was just a single hot and condensed point called singularity, containing all the matter in the Universe. In addition, all four fundamental forces (the gravitational, electromagnetic, strong and weak forces) were unified as a single force.

Stephen Hawking has proposed a theory on what was there before the Big Bang, will leave it for you to Google and find out if interested!

Look Up Question 3
What is Space?

Figure: Picture of empty space as described by quantum physics.

In astronomical sense, space is everything in the Universe above the Earth's atmosphere – the Moon, area where the GPS satellites orbit, other stars, the Milky Way, black holes etc. Space also means whatever area is between planets, Moons, stars, etc. – it's the near-vacuum otherwise known as the interplanetary and the interstellar medium. Space is usually regarded as being completely empty. But this is not true. The vast gaps between the stars and planets are filled with huge amounts of thinly spread gas and dust. Even the emptiest parts of space contain at least a few hundred atoms or molecules per cubic meter.

Space is also filled with many forms of radiation that are dangerous to the astronauts. Much of these infrared and ultraviolet radiations come from the Sun. High energy X-rays, gamma rays and cosmic rays and particles travelling close to the speed of light, arrive from the distant star systems.

Einstein had defined space in a different way by introducing concept of space time that I will explain in a different Look Up question later.

But all this explanation is for classical definition of space. Per quantum physics, empty space is full of quantum fluctuations. That gives rise to umpteen possibilities that we are yet to discover.

Look Up Question 4
What is time and when did it start?
What is Planck time and Planck length?

Based on theory of the Big Bang, there was no time before birth of the Universe. Time came into existence only during the Big Bang.

In classical definition, time is an observed phenomenon, by means of which human beings sense and record changes in the environment and in the Universe. Time has been called an illusion, a dimension, a smooth-flowing continuum, and an expression of separation among events that occur in the same physical location.

Physicists define time as the progression of events from the past to the present and then into the future. Basically, if a system is unchanging, it is timeless. Time can be considered the fourth dimension of reality, used to describe events in three-dimensional space. It is not something we can see, touch, or taste, but we can measure its passage.

Physics equations work equally well whether time is moving forward into the future (positive time) or backward into the past (negative time). However, time in the natural world has one direction, called the arrow of time. The question of why time is irreversible is one of the biggest unresolved questions in science.

Planck unit of time is, how long it takes light to travel one Planck length. So, to find out Planck time we must first know Planck length. The Planck length is the theoretical length at which the laws of space-time begin to unravel. It's orders of magnitude are smaller than a proton, and unable to be physically measured by any equipment available today. The Planck length is approximately 1.6×10^{-35} meters. With this small Planck length, the Planck time is calculated to be around 10^{-43} seconds.

Planck units are used to describe time where we need very small intervals. Example they are used to describe the first few seconds of the Big Bang.

Look Up Question 5
What is the actual shape of our Universe?

Thinking about the shape of the Universe is a bit absurd. When you consider the shape of anything, you view it from outside – yet how could you view the Universe from outside?

Most people would expect the Universe in its entirety to be symmetrically round – a sphere-like shape, sort of like the Earth. Speaking of the Earth, let's consider it for a while. We know the Earth is not flat, but what does that mean? Geometrically, it means that parallel longitudinal lines on its surface aren't parallel. All lines, even if they do start parallel, would end up uniting at one of the Poles, and the distance between them will not be constant.

So, a flat Universe would mean that drawn parallel lines remain parallel – and herein lies the key. Broadly speaking and simplifying things, scientists noted that the light from several galaxies remains parallel to each other, across large distances of the Universe. This pushes us to think that the Universe is indeed flat, and the implications are puzzling. While we can't say with certainty that it will never come to an end – it's likely the Universe extends forever in the space and will go on forever in time. Our results are consistent with an infinite Universe.

Finite vs. Infinite: Another debate about the Universe that involves its shape is its infinite nature. The Universe is big, we know that, but is it really infinite?

Since ancient times, humans have asked themselves that question. In the 18th century, German astronomer Heinrich Wilhelm Olbers came up with a paradox: if the Universe is infinite, then the sky wouldn't be dark. Why? Because in any direction you'd look, there would be infinite space and eventually, you'd encounter a star which would send its light. The night sky doesn't completely light up so voila, the Universe isn't infinite. Unfortunately, it's not as simple as that. There could be a number of explanations for Olber's paradox, and none are simple. The Universe is expanding rapidly, so distant stars are red-shifted into obscurity. Or light from other stars simply hasn't reached us yet – again, we can't really know. What we do know is that objects more than about 13.7 billion years old (the latest figure) are too far away for their light ever to reach us.

We pretty much know the age of the Universe, but what about its size?

The diameter of the observable Universe is 91 billion light-years. The distance the light from the edge of the observable Universe has travelled is very close to the age of the Universe times the speed of light, 13.8 billion light-years, but this does not represent the distance at any given time because the edge of the observable Universe and the Earth have since moved further apart. Because we cannot observe space beyond the edge of the observable Universe, we can't know directly whether the Universe is infinite or not. Modern measurements, including those from the Cosmic Background Explorer (COBE), Wilkinson Microwave Anisotropy Probe (WMAP), and Planck maps of the CMB, suggest that the Universe is infinite in extent, but it's still an ongoing debate.

Most people have a basic idea about what happened during the Big Bang, and this is where the most misconceptions lie. Because that's when space came into existence, most people imagine it expanding in all directions equally – but that's likely not how things went down. Before the Big Bang, there was no space or time. So, there is nothing "outside" the Big Bang in which the Universe could expand to. The Universe simply expanded from a very small volume into a huge volume, and this expansion is occurring even today – but there's no guarantee that the expansion took place symmetrically, in all directions. This asymmetrical expansion means that in most probability shape of the Universe is not sphere.

Let's get back to the local and global geometry. What our scientific observations are detecting is the local geometry, the observable Universe (we can only observe the "observable" – hence the name). Our observations, as thorough and well thought as they may be, can't be exhaustive.

The overall shape of the Universe: So, in the end, we're left with a potentially flat, potentially infinite Universe, but what is its global shape? Unfortunately, we don't really know. Even if lines are parallel and the observable Universe is flat, it doesn't mean that the whole Universe is flat. It could be a Möbius strip for all we know — a shape where space bends and distorts, but lines stay parallel, ultimately connecting one end of space to another.

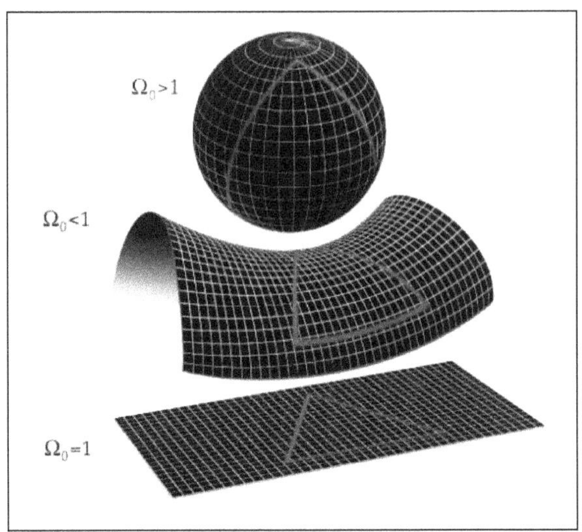

Figure: What could be shape of our Universe?

At the end of the day, there are three distinct possibilities, all with their own distinct implications:

- Universe with zero curvature. A flat Universe. Not necessarily infinite, and not necessarily looking like a sheet of paper. It could also have a shape like a torus.
- Universe with positive curvature. A sphere-like shape.
- Universe with negative curvature. You can think of it locally as a three-dimensional analogue of an infinitely extended saddle shape.

We don't know which one is true. What we do know is that the observable Universe is flat – and that in itself is a remarkable finding.

Look Up Question 6
What happened during first second of the Big Bang? What were the building blocks of the baby Universe?

We looked at what is Universe and how it was created in first question. Now to understand the full evolution of the Universe we must divide the entire time of 13.7 billion years in to following phases:

- Singularity
- Inflation
- Primordial Soup
- Recombination
- Dark Ages
- Birth of Stars and Galaxies
- Birth of our Sun and Solar system
- Planets formation

In this question, we will look at Singularity and Inflation phases. Other phases are explained in question #12.

Singularity

In the beginning, there was nothing. No matter, no energy, no time not even empty space, as the space itself did not exist. Everything was compressed into a "primitive egg" or you can call it a tiny dot that was even smaller than a subatomic particle. The four forces that govern the world of physics were unified into a single symmetry (super force), then from nowhere appeared a fireball smaller than an atom. 10 trillion, trillion times hotter than the core of the Sun. Everything that makes the Universe today exploded from a point that was million times smaller than the tip of a pin. With this explosion time began with first tick of Planck time (10^{-43}s).

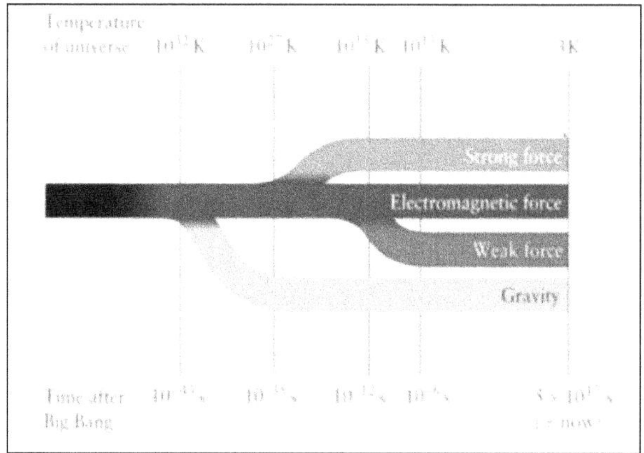

Figure: Sequence in which four fundamental forces came into existing during Big Bang.

During the second tick of Planck time, the tiny fireball Universe expanded and the super force started to break apart. Gravity was the first one to split. At 10^{-35}seconds on the Planck time, strong nuclear force broke apart then came the weak nuclear and electromagnetic force at 10^{-12}seconds. These four fragments of forces are still at work today. Gravity keeps the Earth in orbit. Nuclear forces make Sun and stars shine and electromagnetic radiations (light) gives energy to our planet.

With the start of first Planck tick (at 10^{-43}s), space also came into existence. The Universe, at this time, was very tiny and was a hot mixture of photons, quarks and antiquarks in equilibrium. Due to very high temperature at 10^{32}K, quark and antiquarks iolated together and formed photons. Photons inturn collided with each other and formed quark and antiquark pair again. The Universe was like hot soup of energy and matter changing its form back and forth in the existence gravity (separated from super force at the same time).

The Inflation

At 10^{-35}s Planck time and temperature around at 10^{27}K, strong nuclear force separated rapidly. This rapid separation of strong force triggered rapid expansion of space, in very small span of time. This rapid growth of space was so fast, it is called Inflation. Universe grew exponentially by factor of approximately 10^{26} between 10^{-36}s and at 10^{-32}s from the Big Bang. Inflation phase is critical phase in the Universe creation as it explains why the Universe is nearly flat for all geometrical and calculation purposes and why Universe has structures thatgave birth to stars, galaxies and then us. We will look at the Inflation phase in more detail in a later question.

At 10^{-12}s and temperature around at 10^{27}K, electromagnetic and weak nuclear forces separated. All four forces were separated from this point and are so till today. These four forces started working in their scale and manifestation from this time till now. The Universe at this point was still very hot and was filled with quark-gluon plasma, we can say matter was started to come into being. Also at this phase transition, the Higgs boson particle appeared. What this means is that all the fundamental particles gained mass from this point moving forward in the Universe. We can relate this period to Einstein's famous equation $E=mc^2$, where the energy started converting in to mass.

At 10^{-6} s, and temperature around at 10^{13}K, free quarks formed hadrons, mostly protons and neutrons. Free quarks started to bond together as we see them today in to protons and neutrons. Matter as we understand it now began to emerge. Free neutrons were stable during this time.

At 1 sec and temperature around at 10^{10}K, neutrinos decoupled from matter and streamed freely into the space. Free neutrons were no longer stable. The free neutrinos formed a cosmic neutrino background, which is yet to be observed in the CMB.

So, in summary at the end of one second following had already happened in our Universe:

- It expanded by a factor of 10^{26}
- Super force fragmented and formed four forces of nature
- Inflation gave rise to flat Universe with blue print of its structures.
- Ingredients of matter, elementary particles (like proton, neutrons, and neutrinos) were formed, along with Higgs boson particle and they got mass.
- Matter and antimatter of elementary particles battle settled with matter domination.

Look Up Question 7
What are the age, the size and the mass of the baby and current Universe?

<u>Current Universe</u>

Our Universe is around 13.7 billion years old. The current density of matter in the Universe is about 3 x 10^{-30}g/cm³, which means that it is 300 billion billion billion times less dense than water.

Now, the size of the observable Universe is about 14 billion light years, and using the above value of density it gives us a mass (dark + luminous matter) of about 3×10^{55}g.

Baby Universe

If we consider the age of the baby Universe in the first one second, then the size of the Universe was around 10^{20}meters. The huge size in just one second is due to the inflation concept that we dwelled upon in previous question. Mass of the Universe during the Big Bang is not yet known where the best people in the field are still working on! We don't have clear view on it yet.

Look Up Question 8
Is our Universe static or expanding?
Can we see the whole Universe?

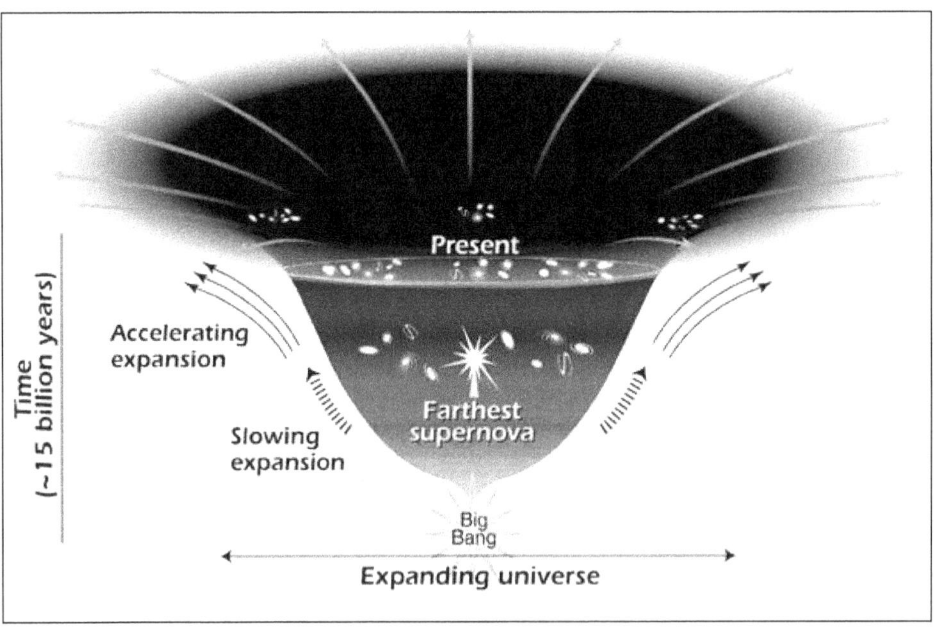

Figure: Expanding Universe.

Till early 20th century, along with Einstein, many other known physicists believed in a static and infinite Universe. Even Einstein had assumed the static

Universe in his famous theory of relativity. But this myth changed when the American astronomer Edwin Hubble made the observations in 1925 and was the first to prove that the Universe is expanding. He observed the distant galaxies and found them speeding away from us. He proved that there is a direct relationship between the speed of distant galaxies and their distances from Earth (This is now known as Hubble's Law). This observation later motivated the theory of the Big Bang that explains the origin of the Universe from small point in space and then its expansion till today.

Look Up Question 9
Where are the centre and the edge of the Universe?

There is no centre of the Universe! According to the standard theories of cosmology, the Universe started with a "Big Bang" about 13.7 billion years ago and has been expanding ever since. Yet there is no centre to the expansion; it is the same everywhere. The Big Bang should not be visualised as an ordinary explosion. The Universe is not expanding out from a centre into space; rather, the whole Universe is expanding and it is doing so equally at all places.

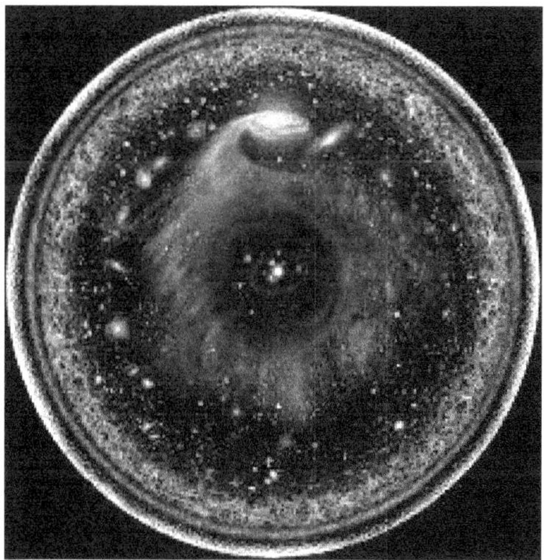

Figure: Observable picture of the Universe on the time dimension.

For an observer, the centre will become his own place when he looks up in all the directions.

Where is "edge of the Universe", what it probably means is "the edge of the visible Universe" or "observable Universe". The oldest light (from the most distant sources) is around 14.5 billion years old. Through a detailed and very careful study of cosmic inflation (as Universe is expanding from the time of its origin) we can estimate that those sources should now be about 46 billion light years away. So, if you define the size of the visible Universe as the present physical distance to the farthest things we can see, then the edge of the visible Universe is 46 billion light years away. However, that "edge" doesn't mean a lot. It's essentially a horizon, in that it only determines how far you can see from the earth.

Look Up Question 10
What is the inflation theory of the Big Bang?

The inflation theory was developed by Alan Guth, Andrei Linde, Paul Steinhardt, and Andy Albrecht, but the evidence for the expansion and the Big Bang theory predate their work.

The Inflation theory proposes that there was a period in the very early stages of the Cosmos (around at 10^{-35}s, after the Big Bang) in which extremely rapid, exponential expansion of the Universe took place prior to the more gradual Big Bang expansion. During the period of inflation, the energy density of the Universe was dominated by a cosmological constant-type of vacuum energy. Later, the vacuum energy decayed to produce the matter and radiation (CMB) we see in our Universe at present. Inflation was rapid. It increased the size of the Universe by a factor of $\sim 10^{26}$ in only a small fraction of a second. When the period of inflation had ceased, essentially, the Universe, that began as a quantum fluctuation about 10^{20} times smaller than a proton, had grown to a sphere about

10 centimetres in diameter within 15 X 10^{-33} seconds. The inflation was rapid enough to overcome gravity and in fact had expanded faster than speed of light.

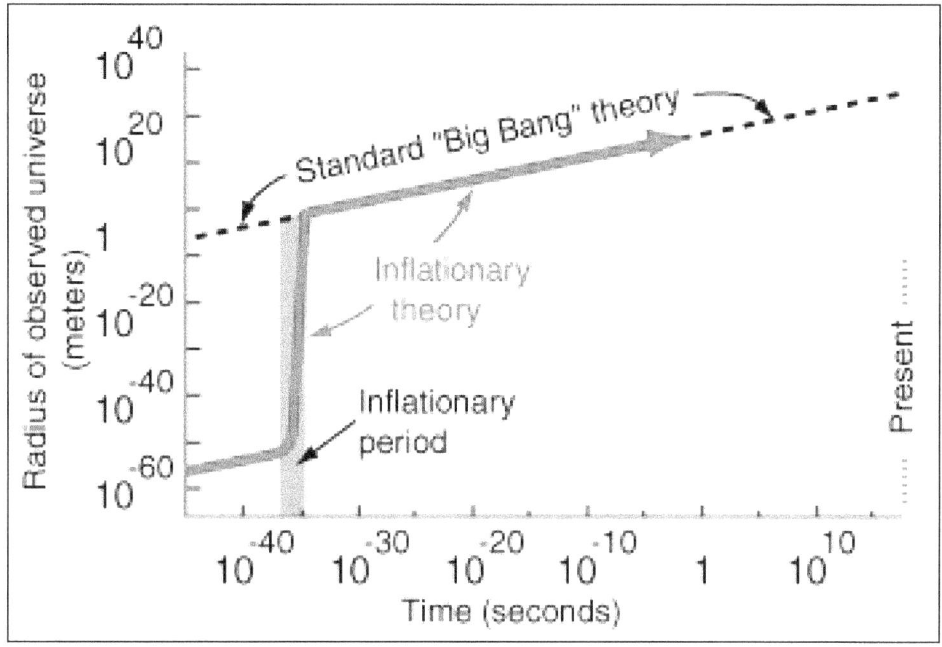

Figure: When inflation happened after the Big Bang.

Inflation provides theoretical answers to some of the problems that appeared in the traditional version of the Big Bang Theory, example - the Horizon Problem and the Flatness Problem:

- **Horizon Problem**

The horizon problem is simply why the Universe looks the same on opposite sides of the sky. The horizon problem may be explained by the rate of initial expansion of the Universe. Initially the horizons of the Universe were in close proximity after which they were separated at a rate exceeding the speed of light.

- **Flatness Problem**

The second problem is that of the Universe's peculiar space-time flat geometry. Inflation offers solutions to this too. The inflation and its subsequent expansion have essentially expanded the curvature of the Universe to such proportions that the human perspective is that of a flat surface. Just as the world appears flat to a casual observer on the surface of the Earth, in a similar way the vast extent of the spherical Universe appears flat to an on-looking astronomer.

Look Up Question 11
How did the baby Universe look like?
What is cosmic wave background (CMB)?

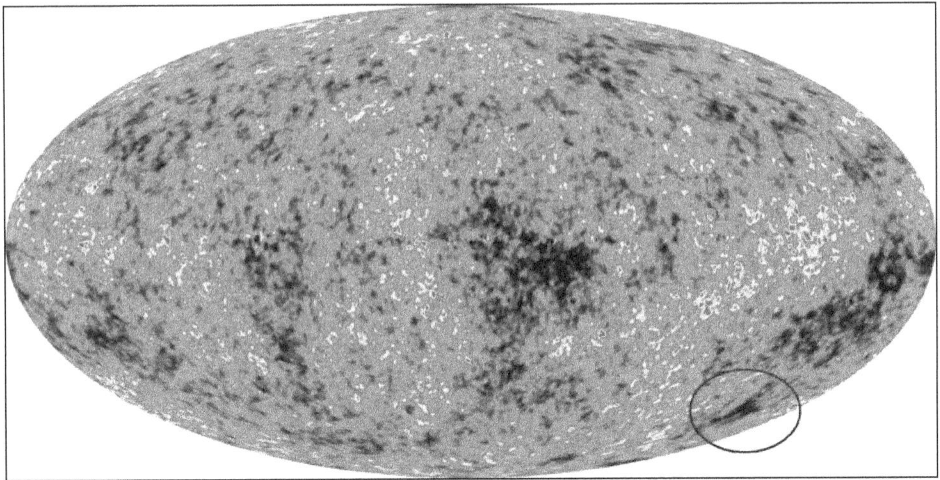

Figure: Cosmic Microwave Background and the cold spot marked by the circle.

The Cosmic Microwave Background radiation, or CMB for short, is a faint glow of light that fills the Universe, falling on Earth from every direction with nearly uniform intensity. It is the residual heat of creation, the afterglow of the big bang.

The CMB is the oldest light we can see. The farthest back both in time and space that we can look, this light set out on its journey more than 13.7 billion years ago, long before the Earth or even our galaxy existed. It is a relic of the Universe's infancy, a time when it was not the cold dark place it is now, but was instead a firestorm of radiation and elementary particles. The familiar objects that surround us today, stars, planets, galaxies eventually coalesced from these particles as the Universe expanded and cooled.

This residual radiation is critical to the study of cosmology because it bears on it the fossil imprint of those particles, a pattern of miniscule intensity variations from which we can decipher the vital statistics of the Universe, like identifying a suspect from his fingerprint.

When this cosmic background light was released billions of years ago, it was as hot and bright as the surface of a star. The expansion of the Universe, however, has stretched the space by a factor of a thousand since then. The wavelength of the light has stretched with it into the microwave part of the electromagnetic spectrum, and the CMB has cooled to its present-day temperature, something the glorified thermometers known as radio telescopes register at about 2.73 degrees above absolute zero.

In 1965 Arno A. Penzias and Robert W. Wilson of Bell Laboratories were testing a sensitive horn antenna which was designed for detecting low levels of microwave radiation. They discovered a low level of microwave background "noise", like the low level of electrical noise which might produce "snow" on a television screen. After unsuccessful attempts to eliminate it, they pointed their antenna to another part of the sky to check whether the "noise" was coming from space, and got the same kind of signal. Being persuaded that the noise was in their instrument, they took other, more sophisticated steps to eliminate the noise, such as cooling their detector to low temperatures.

Finding no explanations for the origin of the noise, they finally concluded that it was indeed coming from space, but that it was the same from all directions. It was a distribution of microwave radiation which matched a blackbody curve for a radiator at about 2.7 Kelvins.

After all their efforts to eliminate the "noise" signal, they found that a group at Princeton had predicted that there would be a residual microwave background radiation left over from the Big Bang and were planning an experiment to try to detect it. Penzias and Wilson were awarded the Nobel Prize in 1978 for their discovery.

What happened after one second of the Big Bang?
What was infant Universe made of? When did the first elementary
particles come into existence?

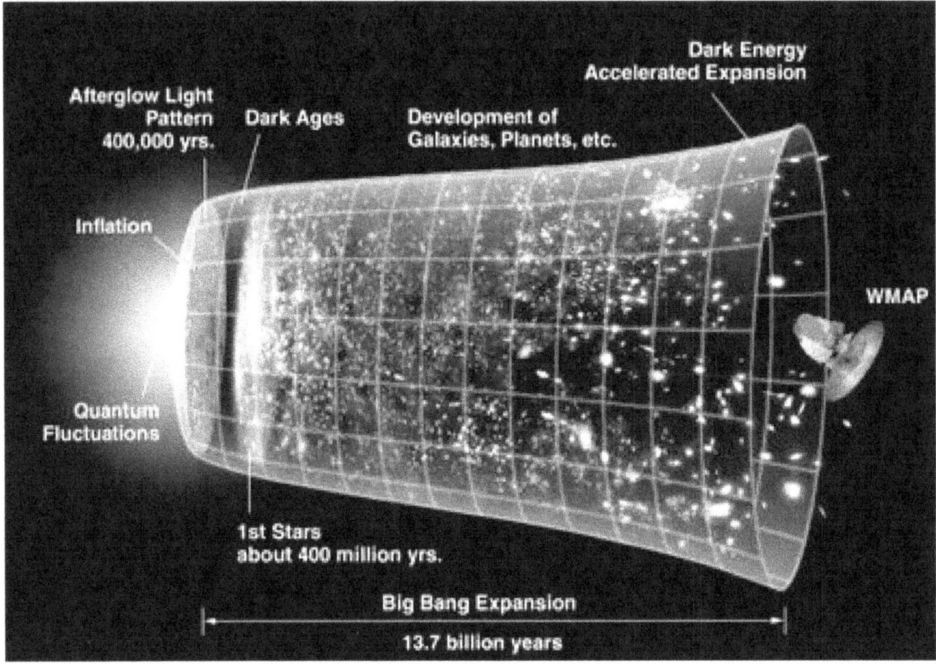

Figure: Evolution of our Universe since the Big Bang.

To understand the full evolution of Universe we can divide the entire time of 13.7 billion years in to following phases:

- Singularity

- The Inflation

- Primordial Soup

- Recombination

- Dark Ages

- Birth of Stars and Galaxies

- Birth of our Sun and Solar system Planets

We have gone through Singularity and Inflation phase in question #10. Out here we will go through rest of the phases:

Primordial Soup:

Primordial soup phase is between 3 to 15 minutes from the Big Bang, when the Universe began to drop down to a billion degrees of temperature. Because this was the time when temperature was cool enough to form nuclei by fusion, the heat was so high that, neutrons, protons and electrons started to join. During this process 23% of Helium, and traces of Deuterium and Lithium were formed that we see today. This phase ended after 12 - 15 minutes because Universe temperature dropped further for fusion to occur (Fusion again happed only after half a billion years when first bunch of stars formed by fusion of Helium in the centre of the stars). At the same time, light photons originated, that were continuously bouncing off or latching onto electrons which caused a constant glow coming from the Universe.

Recombination:

Recombination phase began after 380,000 years from the Big Bang, when temperature dropped to around 3000K; cool enough for charged electrons and nuclei to get recombined into atoms. Formation of electrically neutral Hydrogen and Helium atoms started from here.

The Universe suddenly became transparent with these photons. The cosmic microwave background radiation (CMB)hails from this moment, and thus gives us a direct picture of how matter was distributed at this early time.

Dark ages:

Between Recombination phases from 380000 years till around 500 million years after the Big Bang, the phase that existed is known as Dark Ages. This is the time before first generations of stars were formed. Gas clouds were forming but gravity was not collapsed till the point clouds were dense and cold enough to form the first generations of stars (called Population III stars).

There are many questions that learning more about the dark ages could help answer. For instance, where did the monstrously large black holes seen at the hearts of virtually all large galaxies come from? In addition, a major enigma of the dark ages is how dark matter, the as-yet unidentified material making up about 85 percent of all matter in the Universe, might have influenced the formation of the first galaxies.

Birth of Stars and Galaxies

Star formation acceleration started approximately 500 million years after the Big Bang. At that time, gravity amplified slight irregularities in the density of the primordial gas. Even as the Universe continued to expand rapidly, pockets of gas became more and more dense. Stars ignited within these pockets, and groups of stars became the earliest galaxies.

Birth of the Sun, the Solar System and the Planets

Around 4.5 billion years ago, our Sun was formed within a cloud of gas in a spiral arm of the Milky Way Galaxy. A vast disk of gas and debris that swirled around this new star (Sun) gave birth to planets, Moons, and asteroids we see in our solar system. The planets in our Solar System are believed to have formed from the same spinning disc of dust that formed the Sun. Earth is the third planet from the Sun out of the eight planets we have.

We will also see details of these phases in next set of questions.

Look Up Question 13
Why infant Universe was opaque?

During the first 380,000 years after the Big Bang, the Universe was so hot that all matter existed as plasma. At this time, photons could not travel undisturbed through the plasma because they interacted constantly with the charged electrons and baryons. As a result, the Universe was opaque.

Look Up Question 14
When Universe became transparent?

As the Universe expanded and cooled, electrons began to bind to nuclei, forming atoms. The introduction of neutral matter allowed light to pass freely without scattering. This separation of light and matter is known as decoupling. The light first radiated from this process is what we now see as the Cosmic Microwave Background. The Universe was no longer opaque! However, it would still be some time (perhaps up to a few hundred million years post-Big Bang!) before the first sources of light would start to form, ending the cosmic dark ages. Exactly what the Universe's first light (i.e. stars that fused the existing hydrogen atoms into more helium) looked like, and exactly when these first stars formed is not known.

Look Up Question 15
What is gravity? Why it's considered to be a weak force?
(did you know this?)

Gravity is cosmic glue that binds all the matter in the Universe together.

Gravity is the force that attracts two bodies towards each other, the force that causes apples to fall towards the ground and the planets to orbit the Sun. The more massive an object is, the stronger its gravitational pull. It is what causes objects to have weight. When you weigh yourself, the scale tells you how much gravity is acting on your body. The formula for determining weight is: weight equals mass times gravity. On Earth, gravity is a constant 9.8 meters per second squared, or 9.8 m/s^2.

Sir Isaac Newton developed his Theory of Universal Gravitation in the 1680s. He found that gravity acts on all matter and is a function of both mass and distance. Every object attracts every other object with a force that is proportional to the product of their masses and inversely proportional to the square of the distance between them. The equation is often expressed as:

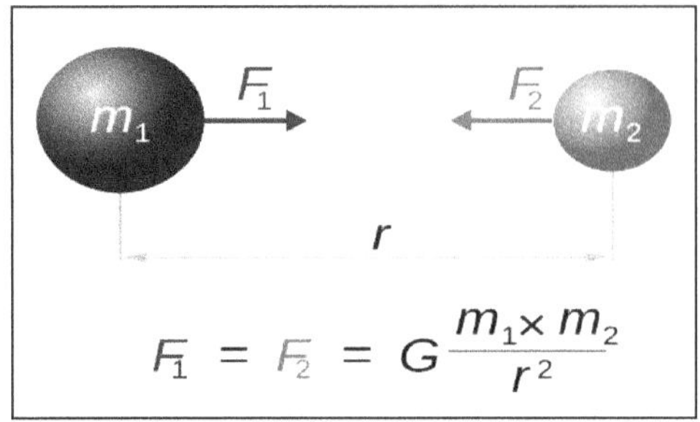

$F_1 = F_2$ is the gravitational force

m_1 and m_2 are the masses of the two objects

r is the distance between the two objects

G is the universal gravitational constant

Newton published his work on gravitation in 1687, which reigned as the best explanation until Einstein came up with his theory of general relativity in 1915. In Einstein's theory, gravity isn't a force, but rather, the consequence of the fact that matter warps space-time. One prediction of general relativity is that light will bend around massive objects.

Physicists see gravity as one of the four fundamental forces that govern the Universe, alongside electromagnetism and the strong and weak nuclear forces. A force is defined as an interaction that changes an object's motion, and so these four forces underpin all of physics and define how everything in the Universe interacts – from the vast cosmic interplay of galaxies to the tight bonds that bind quarks inside a proton or neutron.

Gravity is the weakest of these forces, but it's the one we've been aware of for the earliest times. For centuries, we knew that our feet are kept on the ground and the planets are kept in orbit around the Sun. Even before gravity was described mathematically, seventeenth-century astronomer and mathematician Johannes Kepler had formulated accurate laws to predict the motions of the planets. Still, mysteries remain. Gravity may be one of the fundamental forces of the Universe, but it currently seems fundamentally incompatible with the others. Even before Einstein, physicists strove to formulate a single theory linking every physical aspect of the Universe, including all fundamental forces and particles. A 'theory of everything' would be the ultimate achievement of physics. While quantum field theory (QFT) succeeds in bringing together electromagnetism and the strong and weak nuclear forces, it struggles with including general relativity. Einstein spent the second half of his career trying to puzzle out how gravity fits in, but didn't get very far. Linking these two theoretical frameworks is an open area of research, spanning the field of quantum gravity, string theory, M theory, and more.

Let's see some fun facts around gravity:

- The gravity on the Moon is about 16 percent of that on Earth, Mars has about 38 percent of Earth's pull, while the biggest planet in the solar system, Jupiter, has 2.5 times the gravity of Earth.
- Though nobody "discovered" gravity, legend has it that famous astronomer Galileo Galilei did some of the earliest experiments with gravity, dropping balls off the Tower of Pisa to see how fast they fell.
- Isaac Newton was just 23 years old and back from university when he noticed an apple falling in his garden and began unravelling the mysteries of gravity. (It's probably a myth that the apple bonked him on the head though.)
- An early measure of Einstein's theory of relativity was the bending of starlight near the Sun during a solar eclipse on May 29, 1919.
- Black holes are massive collapsed stars with such strong gravity that even light cannot escape from it.

Although the other forces act over different ranges, and between very different kinds of particles, they seem to have strengths that are roughly comparable with each other. Gravity is the misfit. Why should this be so?

So far, our best explanation comes from string theory, the leading candidate for a "theory of everything". String theory requires that the Universe has more than the three spatial dimensions that we experience possibly as many as 10 dimensions. Per string theorists' best ideas, gravity is so weak because, unlike the other forces, it leaks in and out of these extra dimensions. We only get to experience a dribble of the true strength of gravity.

The proof of this might come through experiments that probe the gravitational attraction between objects that are very small distances apart. String theory suggests that the unseen dimensions are hidden from view because they are rolled up small, or lies "end-on" to our dimensions, making it hard to detect their presence. These compactified dimensions could alter the gravitational attraction between two bodies if they are very small distances apart.

Look Up Question 16
How stars are formed and how they die? When will our Sun die?

Stars are born within the clouds of dust and scattered throughout most galaxies. A familiar example of such as a dust cloud in Milky Way is the Orion Nebula. Turbulence deep within these clouds gives rise to knots with sufficient mass that the gas and dust can begin to collapse under its own gravitational attraction. As the cloud collapses, the material at the centre begins to heat up. Known as a protostar, it is this hot core at the heart of the collapsing cloud that will one day become a star. Spinning clouds of collapsing gas and dust breaks up into two, three or more blobs; this would explain why the majority of the stars in the galaxies are paired or in groups of multiple stars.

Figure: The Orion Nebula, example of stellar in our galaxy Milky Way.

A star the size of our Sun requires about 50 million years maturing from the beginning of the collapse to adulthood. Our Sun will stay in this mature phase for approximately 10 billion years. Stars are fuelled by the nuclear fusion of hydrogen to form helium deep in their interiors. The outflow of energy from the central regions of the star provides the pressure necessary to keep the star from collapsing under its own weight, and the energy by which it shines.

Most stars take millions of years to die. When a star like the Sun has burnt all its hydrogen fuel, it expands to become a red giant. This may be millions of kilometres across - big enough to swallow the planets Mercury and Venus.

After puffing off its outer layers, the star collapses to form a very dense white dwarf. One teaspoon of material from a white dwarf would weigh up to 100 tones. Over billions of years, the white dwarf cools and becomes invisible.

Stars heavier than eight times the mass of the Sun end their lives very suddenly. When they run out of fuel, they swell into red super giants. They try to keep alive by burning different fuels, but this only works for a few million years. Then they blow themselves apart in a huge supernova explosion.

The Sun is currently classified as a "main sequence" star. This means that it is in the most stable part of its life, converting the hydrogen present in its core into helium. For a star the size of ours, this phase lasts a little over 8 billion years. Our solar system is just over 4.5 billion years old, so the Sun is slightly more than halfway through its stable lifetime.

Look Up Question 17
What are different types of stars?

There are many different types of stars, ranging from tiny brown dwarfs to red and blue super giants. There are even more bizarre kinds of stars, like neutron stars. Let's look at all the different types of stars one by one.

Protostar

Figure: The protostar.

A protostar is what you have before a star forms. A protostar is a collection of gas that has collapsed down from a giant molecular cloud. The protostar phase of stellar evolution lasts about 100,000 years. Over time, gravity and pressure increase, forcing the protostar to collapse down. All the energy release by the protostar comes from the heating caused by the gravitational energy only as the nuclear fusion reactions haven't started yet.

T Tauri Star

A T Tauri star is stage in a star's formation and evolution right before it becomes a main sequence star. This phase occurs at the end of the protostar phase, when the gravitational pressure holding the star together is the source of all its energy. T Tauri stars don't have enough pressure and temperature at their cores to generate nuclear fusion, but they do resemble main sequence stars; they're about the same temperature but brighter because they're larger. T Tauri stars can have large areas of Sunspot coverage, and have intense X-ray flares and extremely powerful stellar winds. Stars will remain in the T Tauri stage for about 100 million years.

Main Sequence Star

Figure: Real image of Sun (the main sequence star)

The majority of all stars in our galaxy, and even the Universe, are main sequence stars. Our Sun is a main sequence star, and so are our nearest neighbours, Sirius and Alpha Centauri A. Main sequence stars can vary in size, mass and brightness, but they're all doing the same thing: converting hydrogen into helium in their cores, releasing a tremendous amount of energy.

A star in the main sequence is in a state of hydrostatic equilibrium. Gravity is pulling the star inward, and the light pressure from all the fusion reactions in the star are pushing outward. The inward and outward forces balance one another out, and the star maintains a spherical shape. Stars in the main sequence will have a size that depends on their mass, which defines the amount of gravity pulling them inward.

The lower mass limit for a main sequence star is about 0.08 times the mass of the Sun, or 80 times the mass of Jupiter. This is the minimum amount of gravitational pressure you need to ignite fusion in the core. Stars can theoretically grow to more than 100 times the mass of the Sun.

Red Giant Star:

When a star has consumed its stock of hydrogen in its core, fusion stops and the star no longer generates an outward pressure to counteract the inward pressure pulling it together. A shell of hydrogen around the core ignites continuing the life of the star, but causes it to increase in size dramatically. The aging star becomes a red giant star, and can be 100 times larger than it was in its main sequence phase. When this hydrogen fuel is used up, further shells of helium and even heavier elements can be consumed in fusion reactions. The red giant phase of a star's life will only last a few hundred million years before it runs out of fuel completely and becomes a white dwarf.

Red Dwarf Star

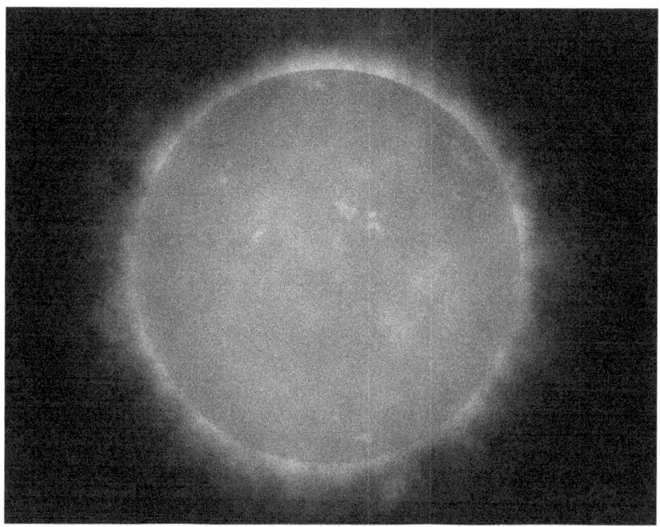

Figure: Red dwarf star

Red dwarf stars are the most common kind of stars in the Universe. These are main sequence stars but they have such low mass that they're much cooler than stars like our Sun. They have another advantage, the Red dwarf stars are able to keep the hydrogen fuel mixing into their core, and so they can conserve their fuel for much longer than other stars. Astronomers estimate that some red dwarf stars will burn for up to 10 trillion years. The smallest red dwarfs are 0.075 times the mass of the Sun, and they can have a mass of up to half of the Sun.

White Dwarf Star

A low or medium mass star (with mass less than about 8 times the mass of our Sun) will become a white dwarf. A typical white dwarf is about as massive as the Sun, yet only slightly bigger than the Earth. This makes white dwarfs one of the densest forms of matter, surpassed only by neutron stars and black holes. When these types of star completely runs out of hydrogen fuel in its core and it lacks the mass to force higher elements into fusion reaction, it becomes a white dwarf star. The outward light pressure from the fusion reaction stops and the star collapses inward under its own gravity. A white dwarf shines because it was a hot star once, but there's no fusion reactions happening any more.

Neutron Stars:

If a star has between 1.35 and 2.1 times the mass of the Sun, it doesn't form a white dwarf when it dies. Instead, the star dies in a catastrophic supernova explosion, and the remaining core becomes a neutron star. As its name implies, a neutron star is an exotic type of star that is composed entirely of neutrons. This is because the intense gravity of the neutron star crushes protons and electrons together to form neutrons.

Figure: Neutron Star.

If stars are even more massive, they will become black holes instead of neutron stars after the supernova goes off.

Supergiant Stars:

The largest stars in the Universe are supergiant stars. These are monsters with dozens of times the mass of the Sun. Unlike a relatively stable star like the Sun, super giants are consuming hydrogen fuel at an enormous rate and will consume all the fuel in their cores within just a few million years. Supergiant stars live fast and die young, detonating as supernovae; completely disintegrating themselves in the process.

A nebula is mostly a cloud of gas and dust in space, and if you have more than one, they are called nebulae. Some nebulae come from the gas and dust thrown out by the explosion of a dying star, such as a supernova. Other nebulae are regions where new stars are beginning to form. For this reason, some nebulae are called "star nurseries."

Nebulae are made of dust and gases — mostly hydrogen and helium. The dust and gases in a nebula are very spread out, but gravity can slowly begin to pull together clumps of dust and gas. As these clumps get bigger and bigger, their gravity gets stronger and stronger. Eventually, the clump of dust and gas gets so big that it collapses from its own gravity. The collapse causes the material at the centre of the cloud to heat up-and this hot core is the beginning of a star.

Nebulae exist in the space between the stars — also known as interstellar space. The closest known nebula to Earth is called the Helix Nebula. It is the remnant of a dying star — possibly one like the Sun. It is approximately 700 light years away from Earth.

There are mainly four main types of Nebulae:

Planetary Nebulae

These are called such because when astronomer William Herschel viewed them through his small telescope in the 1780s he thought their shapes resembled the gas giant planets of our solar system, such as Uranus which he had recently discovered. One of the best examples of a planetary nebula is the Ring Nebula (M57) in Lyra, with other beautiful examples including the Helix Nebula in Aquarius, and the Eskimo Nebula in the constellation of Gemini.

Figure: Planetary Nebula NGC 6369 | by Hubble Heritage

Reflection Nebulae

A reflection nebula does not emit any visible light of its own, and shines only because the light from an embedded source illuminates its dust. The brightest reflection nebulae are stellar nurseries in which its thick gas and dust layers are illuminated by the light being radiated by its young, bright stars, and are often blue in colour. The Trifid Nebula (M20) in Sagittarius is a good example of a reflection nebula illuminated by a group of stars, while another example is the Pleiades reflection nebula in Taurus.

Figure: Messier 78 reflection nebula in Orion

Emission Nebulae

In the Crab Nebula, the Orion Nebula (M42) and other similar emission nebulae, ultraviolet light from young, hot stars strip electrons from the surrounding hydrogen atoms. As they recombine they emit longer wavelengths in the red part of the spectrum, which gives them their distinctive red colour. These types of nebulae often appear to have dark areas as their clouds of dust block the light. Many nebulae have components of both reflection and emission, including the Trifid Nebula.

Figure: Famous Orion emission nebula.

Absorption Nebulae

This type nebula absorbs or obscures the light coming from sources behind them, including stars and bright nebulae, a famous example of which is the Horsehead Nebula. The darkness of the Horsehead is caused mostly by thick dust, although the lower area of the Horsehead's neck casts a shadow to the left. Also known as Dark Nebula, the shapes they form are very irregular, with the largest varieties visible to the naked eye as dark patches, such as the Coalsack Nebula which obscures parts of the bright Milky Way.

Figure: Coalsack Nebula

Look Up Question 19
Why some stars twinkle and some don't?
How far is closest star from Earth?

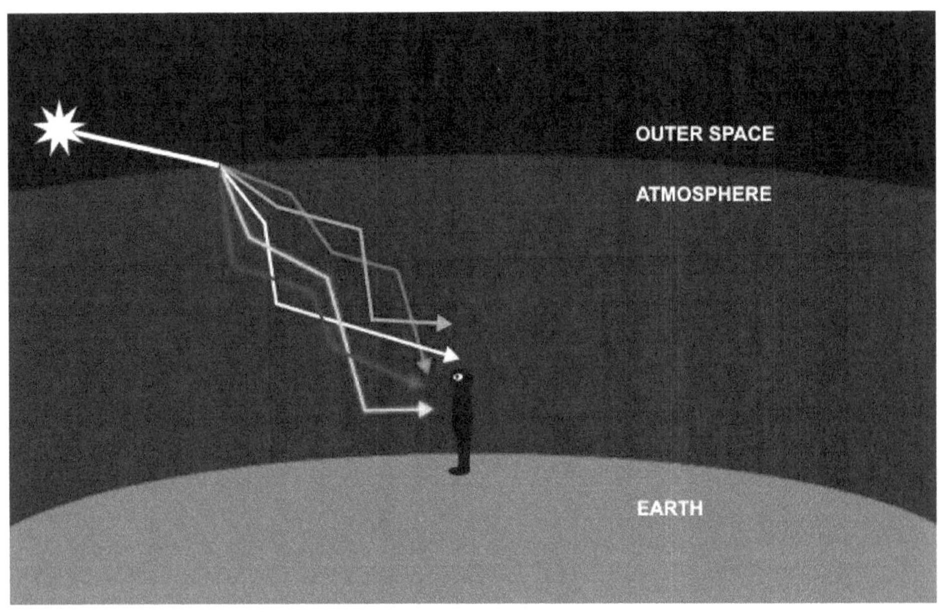

Figure: Why stars twinkle?

Light from other stars travel a very long distance to reach us and passes through pockets of Earth's atmosphere, which vary in temperature and density. Our atmosphere is very turbulent, with streams and eddies forming, churning around, and dispersing all the time. Every layer of Earth's atmosphere has air moving in different directions at different intensities. When light from stars passes through the atmosphere, it is bent due to refraction, which is why stars seem to twinkle when we stare at them. If viewed from outer space, you would not see the stars twinkling.

Some stars don't twinkle even when observing from Earth, because they are planets. Planets are also closer to the Earth than those distant stars, so planets appear larger in comparison. Due to the planets' closeness to Earth, the light coming from these celestial bodies does not bend much due to Earth's atmosphere. Therefore, the light coming from our solar system's planets does not appear to twinkle like stars.

By the way, Sun (our star!) doesn't twinkle, because it's too close to the Earth, as compared to other stars. Due to this, unlike stars, the Sun appears much bigger than a small point in the sky, and therefore, doesn't seem to twinkle.

The Alpha Centauri system is the closest set of stars from Earth (of curse Sun is the closest!). It contains three stars all doing a complex orbital dance together. The primary stars in the system, Alpha Centauri A, and Alpha Centauri B are about 4.37 light-years from Earth. A third star, Proxima Centauri (sometimes called Alpha Centauri C) is gravitationally associated with the former. It's slightly closer to Earth at 4.24 light-years away.

Look Up Question 20
What are constellations?
What are star clusters and binary starsystem?

Constellations:

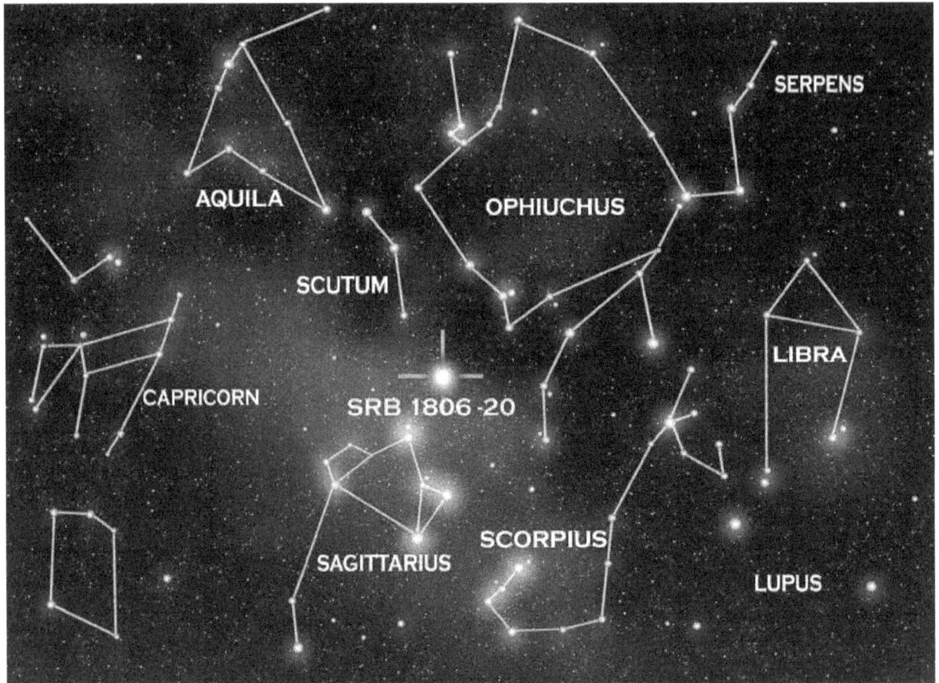

Figure: Examples of star constellations.

Constellations are patterns in the night sky often formed by the most prominent stars that are visible to the naked eye. Technically a constellation defines not just the group of stars that form their patterns but also the region of sky in which it rests.

There are 88 constellations across the sky between the northern and southern hemispheres and, in both these parts of the celestial sphere, these patterns of stars differ. The current list, which includes constellations such as Orion, Cassiopeia, Taurus and the Plough, has been recognized by the International Astronomical Union (IAU) since around 1922 and are based on the 48 which were previously identified by Greek astronomer Claudius Ptolemy.

Constellations often carry names and take the shape of gods, hunters, princesses, objects and mythical beasts associated with Greek mythology – however, at times, it requires quite an imagination to draw out what some constellations are supposed to represent! Some of the most obvious stars in a constellation are often given names and in general, the most visible stars of each constellation are assigned Greek letters with the brightest taking on the first letter of the Greek alphabet (alpha), the second brightest taking beta and so on. As an example, the brightest star in Lyra is Vega which is also called alpha Lyrae.

Constellations can be a useful way to help identify positions of the stars in the sky. Constellations have imaginary boundaries formed by "connecting the dots" and all the stars within those boundaries are labelled with the name of that constellation. However, keep in mind that constellations are not real objects; they are just patterns as seen from our observation point on Earth. The patterns we see are for the most part just by chance. The individual stars in a constellation may appear to be very close to each other, but in fact they can be separated by huge distances in space and have no real connection to each other at all.

Star Clusters

Figure: Star cluster NGC 3766

Stars are formed from gas and dust compressed together by various forces. Star clusters are groups of stars. They have gravitational bounding for some time. When stars are born they develop from large clouds of molecular gas. Due to this they form in groups or clusters, since molecular clouds are composed of hundreds of solar masses of material. After the remnant gas is heated and blow away, the stars collect together by gravity. During the exchange of energy between the stars, some stars reach escape velocity from the proto-cluster and become runaway stars. The rest become clusters of stars, gravitationally bound, meaning they will exist as collection orbiting each other forever.

There are two basic categories of star clusters: Globular and Open (aka. Galactic) star clusters. Globular clusters are roughly spherical groupings of stars that range from 10,000 to several million stars packed into regions ranging from 10 to 30 light years across. They commonly consist of very old Population II stars – which are just a few hundred million years younger than the Universe itself – and are mostly yellow and red. Open clusters, on the other hand, are very different. Unlike the spherically distributed globular, open clusters are confined to the galactic plane and are almost always found within the spiral arms of galaxies. They are generally made up of young stars, up to a few tens of millions of years old, with a few rare exceptions that are as old as a few billion years. Open clusters also contain only a few hundred members within a region of up to about 30 light-years. Being much less densely populated than globular clusters, they are much less tightly gravitationally bound, and over time, will become disrupted by the gravity of giant molecular clouds and other clusters.

Binary Star System

Most of these multiple star systems are binary stars systems. Binary stars are two stars that share a gravitational link and simultaneously orbit their common centre of mass.

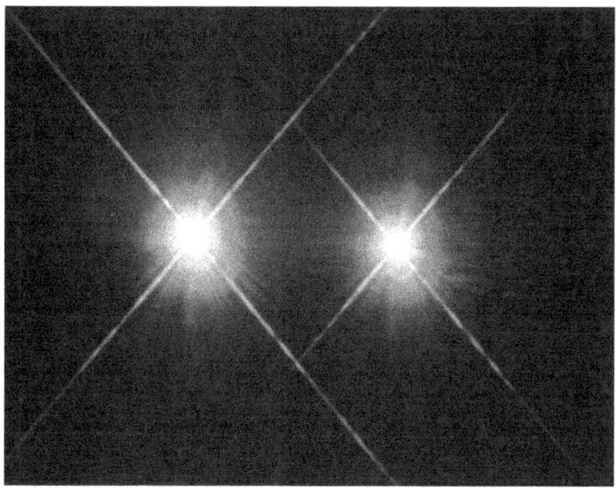

Figure: Alpha Centauri A and B in binary system

In a binary star system, two stars orbit their common centre of mass. Binary stars are classified as either 'wide' or 'close'. In wide binaries, as the name suggests, the orbits of the two stars keep them far apart from each other. The stars move along their life course separately and have little effect on one another. Close binaries, however, are near enough to each other that the gravitational pull of one star can deform and sometimes devour the other star. Since stars are classified based on their masses, this transfer of matter from one star to another can completely alter the stars' life courses.

Look Up Question 21
What are galaxies and galaxy clusters? How they are formed?

Galaxies are huge collections of stars, dust and gas (as well as the invisible, mysterious dark matter), held together by their mutual gravity. They usually contain several million to over a trillion stars and can range in size from a few thousand to several hundred thousand light-years across. Stars are collected together into galaxies. Galaxies are collected together into groups of galaxies, and these groups are collected into clusters. The largest structures in the Universe are galaxy super-clusters that contain millions of galaxies and can measure hundreds of millions of light-years across.

Following the big bang, the primordial Universe consisted of only radiation and subatomic particles. How did it evolve into more than 100 billion galaxies? Scientists have two theories, both of which hinge on the gravitational effects of collapsing gas in the early galaxy.

First, there are the bottom-up theories, in which the gas collapsed and compressed into clumps the size of a million Suns (that's starting small for something the size of the Universe). These clumps then merged to build galaxies. Second the top-down theories, on the other hand, start big. This school of thought argues that the resulting clumps were each the size of multiple galaxies, which in turn broke down into individual galaxies. These latter theories would explain why galaxies occur in clusters.

Either way -- bottom-up or top-down -- the resulting clumps then collapsed into proto-galaxies consisting of dark matter and hydrogen gas. The hydrogen then fell towards the centre of the proto-galaxy while the dark matter remained as an outer halo surrounding it.

Look Up Question 22
How many types and numbers of galaxies are
there in the Universe?

There are hundreds of billions of galaxies in the Universe. As per latest update, there are around 120 billion galaxies in the observable Universe. Galaxies come in many different sizes, shapes and brightness and, like stars, are found alone, in pairs, or in larger groups called clusters. Galaxies are divided into three basic types: spirals, elliptical and irregulars.

Spiral Galaxies

Spiral galaxies are the most common type in the Universe. Our Milky Way is a spiral, as is the rather close-by Andromeda Galaxy. Spirals are large rotating disks of stars and nebulae, surrounded by a shell of dark matter. The central bright region at the core of a galaxy is called the "galactic bulge". Many spirals have a halo of stars and star clusters arrayed above and below the disk.

Figure: How spiral galaxy looks like

Spirals that have large, bright bars of stars and material cutting across their central sections are called "barred spirals". A large majority of galaxies have these bars, and astronomers study them to understand what function they play within the galaxy. In addition to bars, many spirals may also contain supermassive black holes at their cores. Subgroups of spirals are defined by the characteristics of their bulges, spiral arms, and how tightly wound those arms are.

Elliptical Galaxies

Figure: How elliptical galaxy looks like

Elliptical galaxies are roughly egg-shaped (ellipsoidal or ovoid), found largely in galaxy clusters and smaller compact groups. Most elliptical contain older, low-mass stars, and because they lack a great deal of star-making gas and dust clouds, there is little new star formation occurring in them. Elliptical can have as few as a hundred million to perhaps a hundred trillion stars, and they can range in size from a few thousand light-years across to more than a few hundred thousand. Astronomers now suspect that every elliptical has a central supermassive black hole that is related to the mass of the galaxy itself. Messier 87 is an example of an elliptical galaxy. There are some subgroups of elliptical, including "dwarf elliptical" with properties that put them somewhere between regular elliptical and the tightly knit groups of stars called globular clusters.

Irregular Galaxies

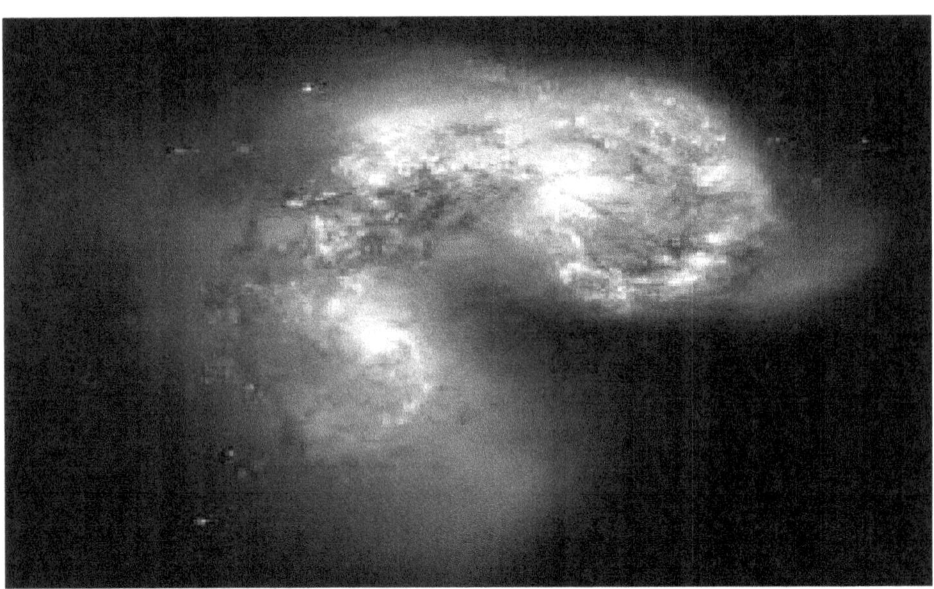

Figure: example of irregular galaxy

Irregular galaxies are as their name suggests: irregular in shape. The best example of an irregular galaxy that can be seen from Earth is the Small Magellanic Cloud. Irregulars usually do not have enough structure to characterize them as spirals or elliptical. They may show some bar structure, they may have active regions of star formation, and some smaller ones are listed as "dwarf irregulars", very similar to the very earliest galaxies that formed about 13.5 billion years ago. Irregulars are characterized by their structures (or lack of them).

Look Up Question 23
What is Milky Way and how big it is? When was it born? Why can't we see its spiral shape from Earth like we can see Andromeda? Where is Earth located in its spiral design?

The Milky Way is our home Galaxy; the spot is where the Sun and the Earth reside. It's a spiral galaxy. Estimated size of Milky Way is around 100,000 light-years across. Since one light year is about 9.5×10^{12} kilo metres, so the diameter of the Milky Way galaxy is about 9.5×10^{17} km in diameter. The thickness of the galaxy ranges depending on how close you are to the centre, but it too is tens of thousands of light-years. We are not anywhere near the centre, we are roughly 165 quadrillion miles from the galaxy's centre (black hole). We live in one of its arms of a large spiral. The Sun and its planets (including Earth) lie in this quiet part of the galaxy, about half way out from the centre. Our Sun and solar system are located in the Orion Arm of Milky Way.

Figure: Our home in Milky Way

Look Up Question 24
Why galaxies collide in expanding Universe?
When will Milky Way merge with Andromeda?

The Universe is expanding but it is still possible that galaxies collide. This is because galaxies that are very close together attract each other gravitationally, counteracting the repulsive force of the expanding Universe.

It is believed that the force of repulsion expanding the Universe is due to the dark energy, i.e. the energy of the vacuum, which is anti-gravitational. This is the force which is pushing the galaxies apart. However, if galaxies are close together, their local gravity can overcome this repulsion. In fact, it is believed that elliptical galaxies are the result of the collision of many smaller galaxies. So, galaxies cannibalise each other all the time, and in the process, create larger elliptical galaxies. Our Milky Way galaxy is on a collision course with its nearest neighbour, Andromeda. Currently, Andromeda and the Milky Way are about 2.5 million light-years apart. Fuelled by gravity, the two galaxies are hurtling toward one another at 402,000 kilometres per hour. But even at that speed, they won't meet for another four billion years. Then, the two galaxies will collide head-on and fly through one another.

Look Up Question 25
Why we find super massive blackholes at the centre of many old and large galaxies?

Figure: Artist's representation of supermassive black hole at the centre of galaxy.

Stellar mass black holes are created when a star at least 5 times larger than the Sun goes out of fuel and collapses in on itself, forming a black hole. The supermassive black holes, on the other hand, can contain hundreds of millions of times the mass of a star like our Sun. The black holes which are found in centre of the galaxy are supermassive black holes.

Astronomers are now fairly certain that these supermassive black holes are at the heart of almost every galaxy in the Universe. Furthermore, the mass of these black holes is somehow tied to the mass of the rest of the galaxy. They grow in tandem with each other.

When large quantities of material fall into the black hole, it chokes up, unable to get consumed all at once. This "accretion disk" begins to heat up and blaze brightly in many different wavelengths, including X-rays. When supermassive black holes are actively feeding, astronomers call these quasars.

So how do these black holes get there in the first place? Astronomers aren't sure, but it could be that the dark matter halo that surrounds every galaxy serves to focus and concentrate material as the galaxy was first forming. Some of this material became the supermassive black hole, while the rest became the stars of the galaxy. It's also possible that the black hole formed first, and collected the rest of the galaxy around it.

Look Up Question 26
How and when our solar system was formed?

Scientists believe that the solar system was formed when a cloud of gas and dust in the space was disturbed, maybe by the explosion of a nearby star (called a supernova). This explosion made waves in space that squeezed the cloud of gas and dust. Squeezing made the cloud start to collapse, as gravity pulled the gas and dust together, forming a solar nebula.

Figure: Our solar system was born 4.7 billion years ago.

Just like a dancer that spins faster as she pulls in her arms, the cloud began to spin as it collapsed. Eventually, the cloud grew hotter and denser in the centre, with a disk of gas and dust surrounding it, that was hot in the centre but cool at the edges. As the disk got thinner and thinner, particles began to stick together and form clumps. Some clumps got bigger, as particles and small clumps stuck to them, eventually forming planets or Moons. Near the centre of the cloud, where planets like Earth formed, only rocky material could stand the great heat. Icy matter settled in the outer regions of the disk along with rocky material, where the giant planets like Jupiter formed.

As the cloud continued to fall in, the centre eventually got so hot that it became a star, the Sun, and blew most of the gas and dust of the new solar system with a strong stellar wind. By studying meteorites, which are thought to be left over from this early phase of the solar system, scientists have found that the solar system is about 4700 million years old!

Look Up Question 27

How and when Moon was born? What are its building blocks?

The Moon has been circling the Earth for more than four billion years. But where did it come from? Some scientists thought that it was captured by the Earth when it came too close. Others think that it was once part of the Earth. Today, most scientists believe it is the 'Earth's child'. It was born when a wandering planet crashed into the young Earth. Huge amounts of material were thrown into space, eventually coming together to form the Moon. This 'big splash' theory would explain why the Moon's rocks are very similar to those on the Earth. Unlike the Earth, the Moon seems to be dead inside.

Today, there are no volcanic eruptions and if any, Moonquakes are very small. At its centre is a small, solid iron core. There is no magnetic field, so explorers could not use compasses on the Moon. The composition of the Moon is a bit of a mystery.

Although we know a lot about what the surface of the Moon is made of, scientists can only guess at what the internal composition of the Moon is. Like the Earth, the Moon has layers. The innermost layer is the lunar core. It only accounts for about 20% of the diameter of the Moon. Scientists think that the lunar core is made of metallic iron, with small amounts of sulphur and nickel. Astronomers know that the core of the Moon is probably at least partly molten. Outside the core is the largest region of the Moon, called the mantle.

The lunar mantle extends up to a distance of only 50 km below the surface of the Moon. Scientists believe that the mantle of the Moon is largely composed of the minerals olivine, orthopyroxene and clinopyroxene.

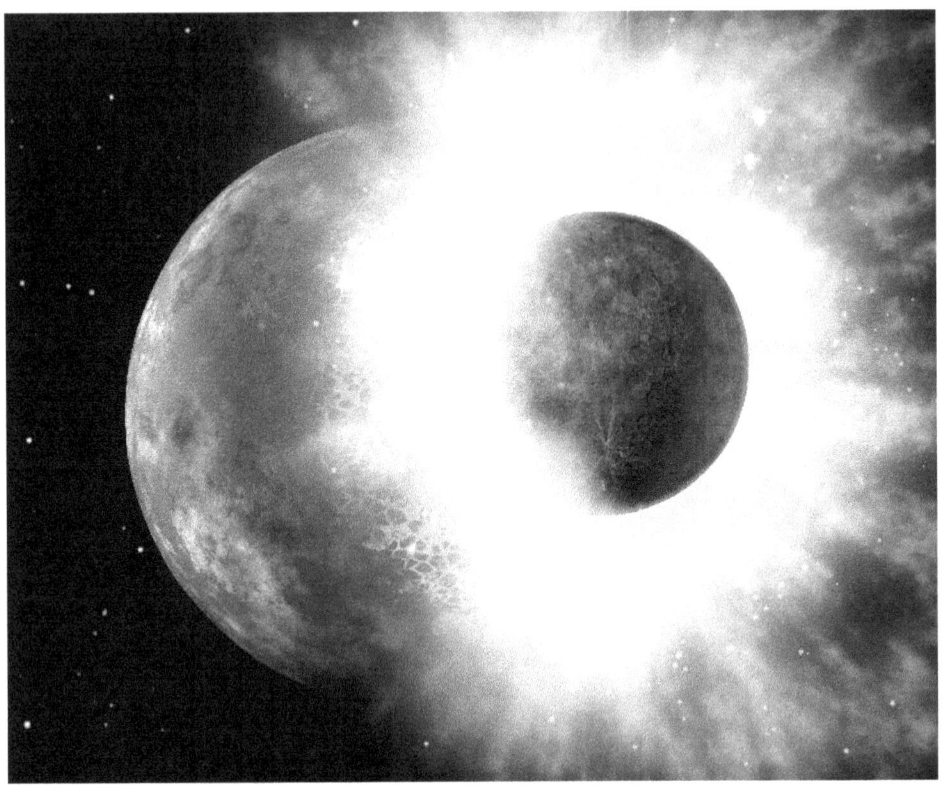

Figure: Big splash theory of moon formation.

It's also believed to be more iron-rich than the Earth's mantle. The outermost layer of the Moon is called the crust, which extends down to a depth of 50 km. This is the layer of the Moon that scientists have gathered the most information about. The crust of the Moon is composed mostly of oxygen, silicon, magnesium, iron, calcium, and aluminium.

There are also trace elements like titanium, uranium, thorium, potassium and hydrogen.

Dark spots on Moon

Figure: Dark spots on moon.

The dark spots on the lunar surface are craters of varying sizes that formed due to collisions from celestial bodies, such as comets, meteorites and asteroids in the past that left large, excavated holes after the impact.

People of every culture have been fascinated by the dark "spots" on the Moon, that seem to compose the figure of a rabbit, frog or the face of a clown or an old lady spinning yarn. With the Apollo missions, scientists found that these features are huge impact basins that were flooded with now-solidified lava.

Dark spots on Sun

Sunspots are darker, cooler areas on the surface of the Sun in a region called the photosphere.

The photosphere has a temperature of 5,800 degrees Kelvin. Sunspots have temperatures of about 3,800 degrees K. They look dark only in comparison with the brighter and hotter regions of the photosphere around them.

Sunspots can be very large, up to 50,000 kilometres in diameter. They are caused by interactions with the Sun's magnetic field which are not fully understood. Sunspots occur over regions of intense magnetic activity, and when that energy is released, solar flares and big storms called coronal mass ejections erupt from Sunspots.

Figure: Sunspots as seen from sun filter.

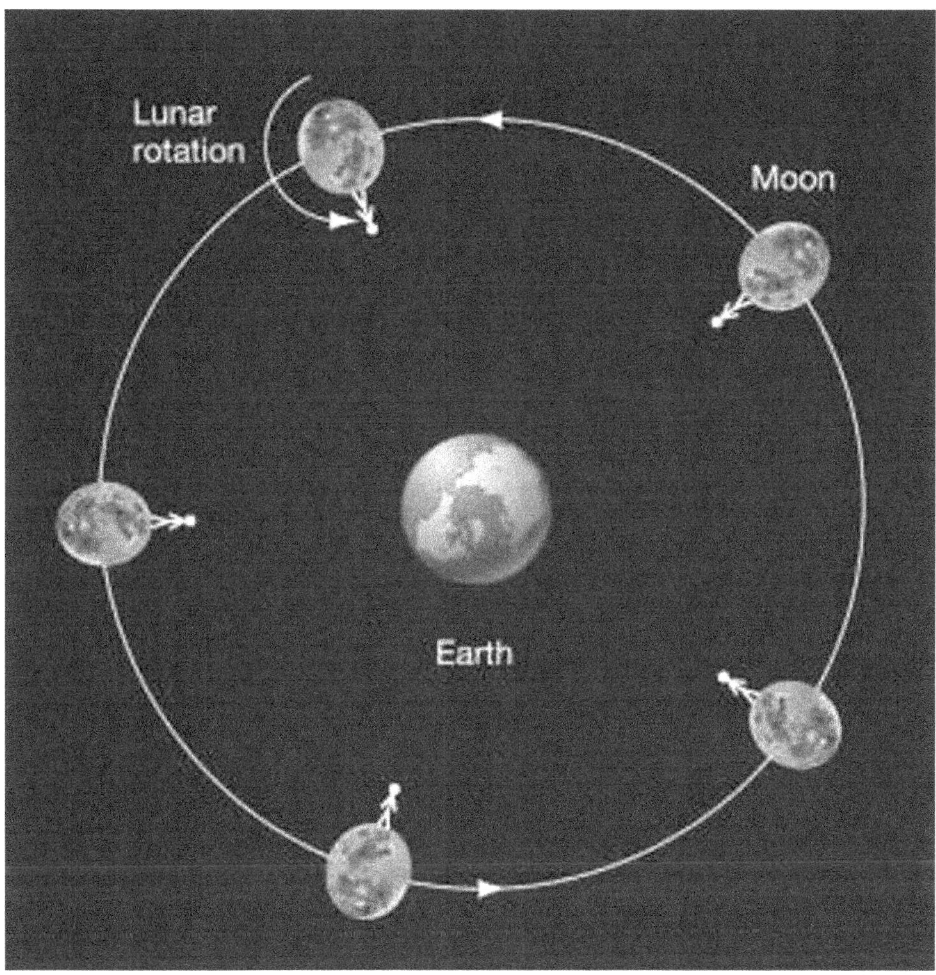

Figure: Moon is tidally locked with Earth.

Moon makes one revolution around the Earth about every 29 days, and that's what causes the Moon phases. But the Moon also rotates once every 29 days. Because of this, the same side of the Moon always faces the Earth:

We call this "being locked." We're not the only system like this, by the way. Both of Mars' Moons, Phobos and Deimos, always have the same side facing Mars. All of Jupiter's, Saturn's, Uranus', and Neptune's Moons are locked to those planets as well. And in a weird case, Pluto and its largest Moon, Charon, are locked to each other, so that both always show one another the same face:

But what made this happen? Why do Moons get locked to the planets they orbit? Like everything else in the solar system, the culprit is gravity. When you stand up, the Earth pulls down on you. But your feet are just a little bit closer to the centre of the Earth than your head. This slight difference means that the Earth pulls with a slightly larger gravitational force on the lower half of you than the upper half; this slight difference is called a tidal force. When something like a Moon is close enough to a planet, these gravitational tidal forces cause the Moon to spin at the same rate that it revolves around the planet. When this happens, we call it tidal locking, and that's why we always see the same side of the Moon!

This is true of all the planets' Moons that we know of, and is even true for some asteroids that are bound to each other. In fact, based on what we know, we think that an extra-solar planet, Gliese 581 c, is tidally locked to its Sun, just like our Moon is locked to us. Pretty neat!

Look Up Question 30
Curious questions on our planets

a)   **What is Mars atmosphere made of?**

b)   **Why Saturn has rings but other planets of solar system don't?**

c)   **Why Jupiter is considered as hot gas giant? Is Jupiter a failed star?**

d)   **Pluto was well recognized last planet of the solar system, why its planet status is removed now?**

Mars atmosphere:

The atmosphere of Mars is less than 1% of Earth's (so it does not protect the planet from the Sun's radiation, nor does it do much to retain heat at the surface). It consists of 95% carbon dioxide, 3% nitrogen, 1.6% argon, and the remainder is trace amounts of oxygen, water vapour, and other gases. Also, it is constantly filled with small particles of dust (mainly iron oxide), which give Mars its reddish hue. Scientist believe that the atmosphere of Mars is so negligible because the planet lost its magnetosphere about 4 billion years ago.

Saturn rings

Saturn has accumulated a great deal of dust, particles, and ice at varying distances from its surface. These items are most likely trapped by gravity. The rings appear because of the wavelengths of light reflected by these rings of debris.

Figure: Spectacular rings of Saturn.

Some scientists speculate that Saturn may be too big. Its gravitational pull is so strong that it has been able to snatch debris from space. Some of which is as large as an entire building. That pull is why it has at least 62 Moons. Those Moons contribute dust to the rings as well as absorb dust from the rings.

A common theory as to how all the material initially accumulated in Saturn's rings is a series of asteroid impacts. Not with the planet, but with the moons around it. After the impact, the remnants of the asteroids and the debris from the moons could not escape the gravitational pull of the planet.

No matter which theory you believe, the rings of Saturn are spectacular. Not only Saturn has rings, there are ring systems around Neptune, Uranus, and Jupiter too. But they are fainter than Saturn's.

Jupiter a gas giant or a failed star

Jupiter, the biggest planet of our solar system is a gas giant along with Neptune and Uranus. Jupiter is called a gas planet because its atmosphere is densely filled with gases. The atmosphere contains 90% hydrogen and 10% helium with small traces of other gases.

Figure: Jupiter and Earth size compared.

It is astonishing that there is no solid surface to stand in Jupiter. First, there is a layer of clouds. Then there are two layers of hydrogen, helium and other gases. And after that there is a layer of liquid metallic hydrogen. Finally, the core appears which is 1.5 times bigger than Earth. The gas planets do not have solid surfaces, their gaseous material simply gets denser with depth.

We can consider Jupiter to look somewhat like a "failed star" because it has a chemical composition that is very similar to that of the Sun. Jupiter is mostly hydrogen and helium, with only a small fraction of any heavier elements. This means that if Jupiter was heavier, then it could undergo fusion (the source of energy in the Sun) and radiate its own light. However, Jupiter is too light, and its central temperature never got high enough for hydrogen fusion to start.

The fact that Jupiter has the same composition as the Sun tells us something interesting about how it formed - that is, it must have formed at the same time as the Sun, out of the same cloud of gas. However, that brings us to the key difference between stars (like the Sun) and planets (like Jupiter). When a star (and its planets) forms, the whole cloud collapses. Due to the conservation of "angular momentum" (i.e., the rate the material is spinning), the collapsing cloud spins up, and some of the material can't fall all the way down to the star. This material instead becomes a disk orbiting around the star, and planets condense out of that disk.

Pluto is not a planet:

Since its discovery in 1930, Pluto has been a bit of a puzzle. It's smaller than any other planet -- even smaller than Earth's Moon. It's dense and rocky, like the terrestrial planets (Mercury, Venus, Earth and Mars). However, its nearest neighbours are the gaseous Jovian planets (Jupiter, Saturn, Uranus and Neptune). For this reason, many scientists believe that Pluto originated elsewhere in space and got caught in the Sun's gravity. Some astronomers once theorised that Pluto used to be one of Neptune's Moons.

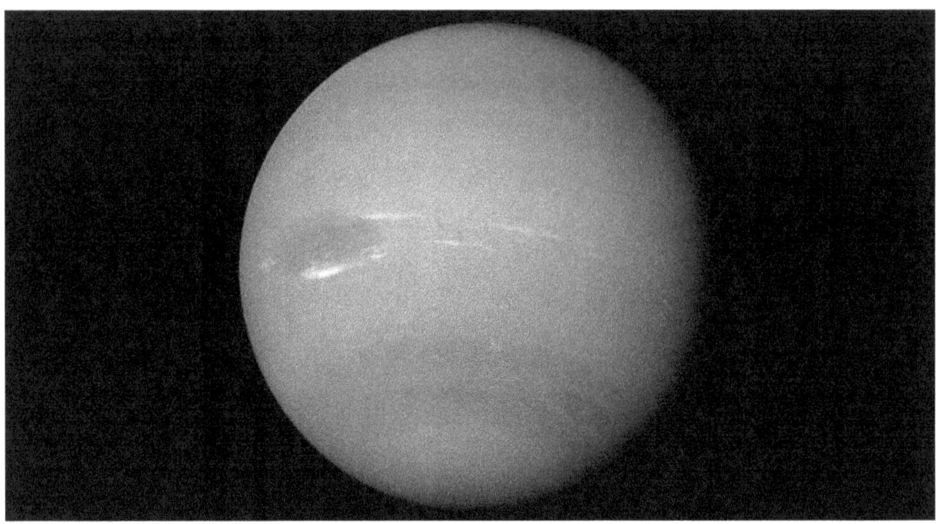

Figure: Pluto is not a planet.

Pluto's orbit is erratic. The planets in our solar system all orbit the Sun in a relatively flat plane. Pluto, however, orbits the Sun at a 17-degree angle to this plane. In addition, its orbit is exceptionally elliptical and crosses Neptune's orbit.

One of its Moons, Charon, is about half Pluto's size. Some astronomers have recommended that the two objects be treated as a binary system rather than a planet and satellite.

These facts contributed to the long-running debate over whether to consider Pluto a planet. On Aug. 24, 2006, the International Astronomical Union (IAU), an organization of professional astronomers, passed two resolutions that collectively revoked Pluto's planetary status. The first of these resolutions was Resolution 5A, that defines the word "planet." Although many people take the definition of "planet" for granted, the field of astronomy had never clearly defined what is and is not a planet.

Here's how Resolution 5A defines a planet:

A planet is a celestial body that (a) is in orbit around the Sun, (b) has sufficient mass for its self-gravity to overcome rigid body forces so that it assumes a hydrostatic equilibrium (nearly round) shape, and (c) has cleared the neighbourhood around its orbit.

Look Up Question 31
What is Einstein theory of relativity and how it had redefined the laws of Universe?

Theory of Relativity – A Brief History

The Theory of Relativity, proposed by the Jewish physicist Albert Einstein (1879-1955) in the early part of the 20th century, is one of the most significant scientific advances of our time. Although the concept of relativity was not introduced by Einstein, his major contribution was the recognition that the speed of light in a vacuum is constant and an absolute physical boundary for motion. This does not have a major impact on a person's day-to-day life since we travel at speeds much slower than light speed. For objects travelling near light speed, however, the theory of relativity states that objects will move slower and shorten in length from the point of view of an observer on Earth. Einstein also derived the famous equation, $E = mc2$, which reveals the equivalence of mass and energy.

When Einstein applied his theory to gravitational fields, he derived the "curved space-time continuum" which depicts the dimensions of space and time as a two-dimensional surface where massive objects create valleys and dips in the surface. This aspect of relativity explained the phenomena of light bending around the Sun, predicted black holes as well as the Cosmic Microwave Background Radiation (CMB)

Theory of Relativity – The Basics

Physicists usually describe the Theory of Relativity into two parts. The first is the Special Theory of Relativity that essentially deals with the question of whether rest and motion are relative or absolute, and with the consequences of Einstein's conjecture that they are relative.

The second is the General Theory of Relativity that primarily applies to particles as they accelerate, particularly due to gravitation, and act as a radical revision of Newton's theory, predicting important new results for fast-moving and/or very massive bodies. The General Theory of Relativity correctly reproduces all validated predictions of Newton's theory, but expands on our understanding of some of the key principles. Newtonian physics had previously hypothesized that gravity operated through empty space, but the theory lacked explanatory power as far as how the distance and mass of a given object could be transmitted through space. General relativity irons out this paradox, for it shows that objects continue to move in a straight line in space-time, but we observe the motion as acceleration because of the curved nature of space-time.

Einstein's theories of both special and general relativity have been confirmed to be accurate to a very high degree over recent years, and the data has been shown to corroborate many key predictions; the most famous being the solar eclipse of 1919 bearing testimony that the light of stars is indeed deflected by the Sun as the light passes near the Sun on its way to Earth. The total solar eclipse allowed astronomers to, for the first time analyze starlight near the edge of the Sun, which had been previously inaccessible to observers due to the intense brightness of the Sun. It also predicted the rate at which two neutron stars orbiting one another will move toward each other. When this phenomenon was first documented, general relativity proved itself accurate to better than a trillionth of a percent precision, thus making it one of the best confirmed principles in all of physics.

Experimental evidence

Although instruments can neither look nor measure space-time, several of the phenomena predicted by its warping have been confirmed.

Gravitational lensing: Light around a massive object, such as a black hole, is bent, causing it to act as a lens for the things that lie behind it. Astronomers routinely use this method to study stars and galaxies behind massive objects.

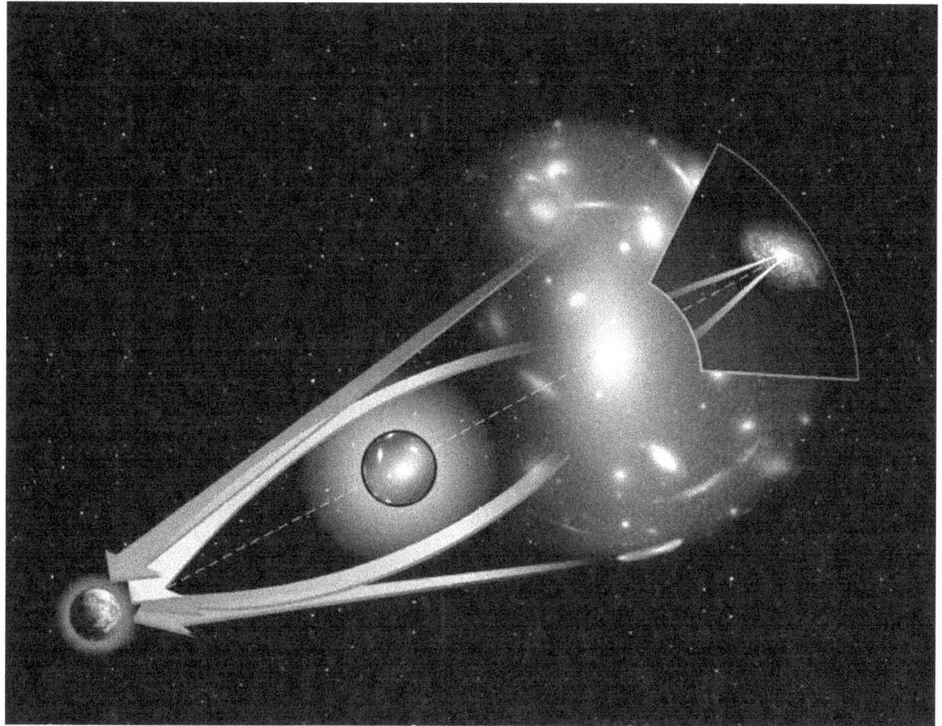

Figure: Gravitational lensing

A quasar in the Pegasus constellation is an excellent example of gravitational lensing. The quasar is about 8 billion light-years from Earth, and sits behind a galaxy that is 400 million light-years away. Four images of the quasar appear around the galaxy because the intense gravity of the galaxy bends the light coming from the quasar.

Gravitational lensing can allow scientists to see some pretty cool things, but until recently, what they spotted around the lens has remained fairly static. However, since the light travelling around the lens takes a different path, each travelling over a different amount of time, scientists could observe a supernova occur four different times as it was magnified by a massive galaxy.

In another interesting observation, NASA's Kepler telescope spotted a dead star, known as a white dwarf, orbiting a red dwarf in a binary system. Although the white dwarf is more massive, it has a far smaller radius than its companion.

Changes in the orbit of Mercury:

The orbit of Mercury is shifting very gradually over time, due to the curvature of space-time around the massive Sun. In a few billion years, it could even collide with Earth.

Gravitational redshift:

The electromagnetic radiation of an object is stretched out slightly inside a gravitational field. Think of the sound waves that emanate from a siren on an emergency vehicle; as the vehicle moves toward an observer, sound waves are compressed, but as it moves away, they are stretched out, or redshifted. Known as the Doppler Effect, the same phenomena occur with waves of light at all frequencies.

Gravitational waves:

Violent events, such as the collision of two black holes, are thought to be able to create ripples in space-time known as gravitational waves. In 2016, the Laser Interferometer Gravitational Wave Observatory (LIGO) announced that it found evidence of these collisions.

Figure: Gravitational waves during collision of two blackholes

LIGO spotted the first confirmed gravitational wave on September 14, 2015. The pair of instruments, based out of Louisiana and Washington, had recently been upgraded, and were in the process of being calibrated before they went online. The first detection was so large that, it took the team several months of analysation to convince them that it was a real signal and not a glitch.

A second signal was spotted on December 26 of the same year, and a third candidate was mentioned along with it. Together, the two firm detections provide evidence for pairs of black holes spiralling inward and colliding. As time passes, it is anticipated that more gravitational waves will be detected by LIGO and other upcoming instruments, such as the one planned in India.

Relativity in summary:

- Prior to Einstein, Issac Newton's laws were used to understand the physics of motion. In 1687, Newton wrote that gravity affects everything in the Universe. The same force of gravity that pulled an apple down from a tree kept the Earth in motion around the Sun. But Newton never explained the source of gravity.

- Philosopher David Hume thinking about time and space left deep influence on Einstein thinking. "It is very well possible that without these philosophical studies I would not have arrived at the solution." Einstein wrote.

- In 1905, Albert Einstein based a new theory on two principles. First, the laws of physics appear the same to all observers. Second, he calculated that speed of light is unchanging. Prior to Einstein, scientists believed that space was filled with aether that would cause the speed of light to change depending on the relative motion of the source and the observer.

- As a result of these principles, Einstein deduced that there is no fixed frame of reference in the Universe. Everything is moving relative to everything else. It is known as special relativity because it applies only to special cases: fames of reference in constant changing motion. In 1915, Einstein published the general theory of relativity, to frame that are accelerating with regard to each other.

- Time does not pass at the same rate for everyone. A fast-moving observer measures time passing more slowly than a (relatively) stationary observer would. The phenomenon is called time dilation.

- A fast-moving object appears shorter along the direction of motion, relative to slow moving one. This effect is very subtle until the objects travel close to speed of light.

- Mass and energy are different manifestation of same thing. Einstein's famous equation $E = mc^2$ we all know. This is what enables the release of a huge amount of energy from a nuclear explosion.

- As a result of $E = mc^2$, a fast-moving object appears to have increased mass relative to slow moving one. This is due to the fact that increasing an objects velocity increases its kinetic energy, and therefore its mass.

- The increase in mass is the reason that Einstein says that matter cannot travel faster than light. The mass increases with velocity until the mass becomes infinite when it reaches light speed. An infinite mass would require energy to move, so this is impossible.

- Space and time are part of one continuum, called space-time. In Einstein mathematics, space has three dimensions, and fourth dimensions are time. Space-time can be thought of as a grid of fabric. The presence of mass distorts space-time, so the rubber sheet model is popular to visualize.

- Relativity explains where gravity comes from. The rubber sheet model shows that gravity results from massive objects warping space-time. The warp is called a gravity well. Orbiting objects follow the path that is shortest and requires least amount of energy. The planets move in ellipses, the most energy efficient part in the gravity well of sun.

- Gravity bends light. This phenomenon is called gravitational lensing. When we observe a distant galaxy, the gravity of the matter between Earth and the galaxy causes light rays to bend into different paths. When the light reaches the telescope, multiple images of the same galaxy appear.

Look Up Question 32
Why speed of light is always constant?

No one knows and explained this question satisfactorily. Let us see few attempts that I could find on this topic.

If you start from Maxwell's equations, which are the set of equations that describe how electric and magnetic fields are related to each other and charge, you will see that these are coupled, i.e. the electric field is described by the change of the magnetic field and vice versa. It is possible to decouple these equations. When you do this with for either of the fields (in vacuum) you end up with a wave equation for the field, which in physics is a well-known type of equation. From this you get the speed of propagation, in this case just the speed of light. So,you could say that the light moves at this speed, because if you solve the necessary equations you end up with this answer.

Now comes the part with special relativity. Already at the end of the 19th century they knew that light should move at this speed, but at this point you operated with Galilean relativity. This is just that if you move with some speed v_1 relative to something stationary (the ground maybe) and another object moving in the same direction, moving with speed v_2 relative to this reference point, then your speed relative to this other object would be v_1-v_2. The assumption here being that you can define some absolute frame of reference that everything moves relative to.

Something weird happens though, people noticed. If you look at the light from something moving away from you, you would think it should be moving slower. If the object was moving towards you, you would think light coming to you would be moving faster. In reality, this doesn't happen. You also get some weird conundrums that make Maxwell's equations break down if you pursue the idea of an absolute frame of reference.

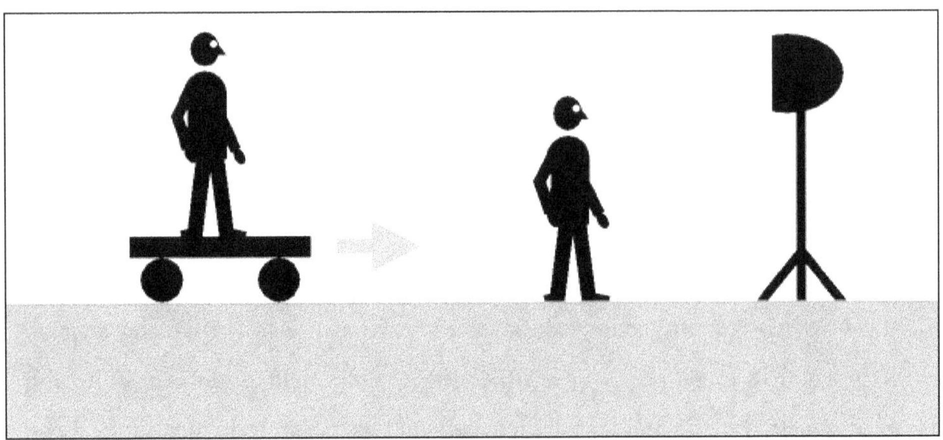

Figure: Speed of light is always constant for any user (moving or stationary).

Einstein's solution to this problem was to say that instead of having an absolute frame of reference (space being constant so to say or time for that matter), the only absolute should be light moving at the speed 'c', no matter what frame of reference we are observing from. This has some weird consequences, namely that space and time changes depending on your frame of reference, in a way that makes light always move at speed 'c'. On the bright side: now Maxwell's equations (and other laws of physics) behave as they should no matter where you view them from. So, in short: the speed of light is fixed because it makes the laws of physics well behaved no matter how you move relative to any frame of reference.

Here is another theory or explanation that Einstein himself never realised (not sure if this is true but found it interesting). The constancy of the speed of light is NOT a fundamental law of nature, because there is a simple underlying explanation for it. And it is an explanation that cannot be avoided. We perceive the speed of light to be constant because the physiological functioning of our body (including our brain) depends on the speed of electromagnetic transmission (which is the speed of light), and we have defined our time according to how we physiologically experience it. We have thus effectively used the speed of light as the standard for defining our time, and that is why the speed of light is always constant per our time. As an example, let us look at how nerve cells function in our body. A nerve cell communicates with another nerve cell through the transmission of electrical impulses along the axon of the cell that connects it to this other nerve cell. The electrical impulse is caused by the movement of charged particles (mainly sodium and potassium ions) along the axon of the nerve cell. The movement of charged particles disturbs the surrounding electrical fields, and this disturbance is transmitted to other charged particles along the axon—that is how the electrical impulse proceeds. Now, the rate at which this disturbance is transmitted is the rate of electromagnetic transmission, which is again the speed of light. That is why the rate at which our nerve cells function depends on the speed of light.

Naturally, the rate at which our nerve cells function also determines the rate at which our brains function. And the rate at which our brains function determines that rate at which we think. This means that the rate at which we think is also dependent on the rate of electromagnetic transmission, which is the speed of light. This same consideration can be applied to all the cells and all the organs of our body. Thus, the rate of electromagnetic transmission, which is the speed of light, determines the rate at which our body functions.

That provides the explanation for the constancy of the speed of light. We are essentially trapped inside the system, like the characters inside a video movie. If someone slows down the video, all the characters inside the video as well as everything else there slow down in same proportion. The characters' movements and even their rate of thinking, slows down proportionally in line with everything else. It is therefore not possible for these characters inside the video to notice any difference in the speed of the video itself. They would not notice any change at all, simply because they are trapped inside the system and cannot escape it to view it from the outside.

Likewise, we cannot measure any change in the speed of light because we are also "trapped inside the system." And we are trapped because the speed of light also determines the rate at which our body functions. If the speed of electromagnetic transmission slows down, we ourselves, our actions and our thoughts slows down by the same amount, so we cannot detect any change. That is why we always perceive the speed of light to be constant. It is not a fundamental law of nature at all; it is merely a consequence of how our body functions!

Thus, it can be seen that relativity exists because our science is actually a science of what we experience, and not a science of a Universe "out there" that is independent of us as observers. This means that the role of the conscious observer in our science must be taken into account.

Look Up Question 33
What are gravitational waves and how they are detected?

Gravitational waves are 'ripples' in the fabric of space-time caused by some of the most violent and energetic processes in the Universe. Albert Einstein predicted the existence of gravitational waves in 1916 in his general theory of relativity. Einstein's mathematics showed that massive accelerating objects (such as neutron stars or black holes orbiting each other) would disrupt space-time in such a way that 'waves' of distorted space would radiate from the source (like the movement of waves away from a stone thrown into a pond). Furthermore, these ripples would travel at the speed of light through the Universe, carrying with them information about their cataclysmic origins, as well as invaluable clues to the nature of gravity itself.

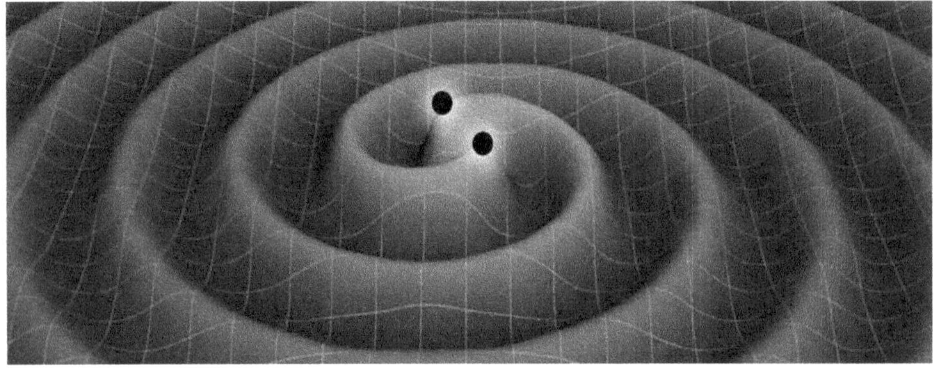

Figure: Gravitation waves during collision of two supermassive black holes.

The strongest gravitational waves are produced by catastrophic events such as colliding black holes, the collapse of stellar cores (supernovae), coalescing neutron stars or white dwarf stars, the slightly wobbly rotation of neutron stars that are not perfect spheres, and the remnants of gravitational radiation created by the birth of the Universe itself.

How do we know that gravitational waves exist?

In 2015, scientists detected gravitational waves for the very first time. They used a very sensitive instrument called LIGO (Laser Interferometer Gravitational-Wave Observatory). These first gravitational waves happened when two black holes crashed into one another. The collision happened 1.3 billion years ago. But, the ripples didn't make it to Earth until 2015!

The first detection of gravitational waves was a very important event in science. Before this, just about everything we knew about the Universe came from studying waves of light. Now we have a new way to learn about the Universe — by studying waves of gravity.

Gravitational waves will help us learn many new things about our Universe. We may also learn more about gravity itself!

How are gravitational waves detected?

When a gravitational wave passes by Earth, it squeezes and stretches space. LIGO can detect this squeezing and stretching. Each LIGO observatory has two "arms" that are each more than 2 miles (4 kilometres) long. A passing gravitational wave causes the length of the arms to change slightly. The observatory uses lasers, mirrors, and extremely sensitive instruments to detect these tiny changes.

What are black holes? How they are formed and
how we detect them?

A black hole is a region of space where absolutely nothing can escape. That's because they have extremely strong gravitational effects, which means once something goes into a black hole, it can't come out. They get their name because even light can't escape once it's been sucked in – which is why a black hole is completely dark.

Figure: Artist representation of supermassive blackhole at the centre of galaxy.

Black holes are places of mystery. The laws of physics predict their existence but cannot explain what happens inside a black hole. Once we can do that, we will have stepped beyond the work of Albert Einstein and taken the next big leap in our understanding of the Universe.

How are black holes created?

Most black holes are made when a supergiant star dies. This happens when stars run out of fuel – like hydrogen – to burn, causing the star to collapse.

When this happens, gravity pulls the centre of the star inwards quickly, and collapses into a tiny ball.

It expands and contracts until one final collapse, causing part of the star to collapse inward thanks to gravity, and the rest of the star to explode outwards.

The remaining central ball is extremely dense, and if it's especially dense, you get a black hole.

Put simply, black holes are places where gravity is so strong that nothing can escape once it gets too close. This sets them apart from all other celestial objects, where you would always – in principle – be able to build a rocket strong enough to escape into space.

The boundary of no return is called the event horizon. Once across this invisible threshold you will be in the black hole's clutches forever. It's the celestial equivalent of a lobster pot – easy to get into but impossible to leave.

The concept of a black hole follows naturally from Albert Einstein's General Theory of Relativity, published in 1915. It states that the strength of a celestial object's gravitational field is determined by the density of matter it contains: the higher the density, the stronger the gravity.

The dilemma is where does this process stop? In a black hole, there is nothing known that can resist the overwhelming gravity, and we are forced to believe that the matter is simply crushed out of existence. This is a dramatic way of describing a mathematical object called a singularity, a point of zero volume and infinite density. But physicists are unsure that such objects can exist in reality. It just sounds impossible.

Theories of quantum gravity are designed to tackle this problem and tell us how gravity behaves on the smallest possible scales, and at the great possible strengths and densities of matter. There are a number of approaches being taken, including the often-mentioned string theory, but at present there is no strong front-runner.

A successful quantum gravity theory would not only allow astronomers to study the interior of a black hole, it would also tell us about the big bang, the origin of the Universe. This is because the big bang is another place in nature where relativity predicts a singularity will appear. In short, a quantum theory of gravity could tell us how the Universe came into existence in the first place, and that's way astronomers are still so interested in black holes.

There are two basic ways we can detect whether there's a black hole. Black holes are, of course, dark by definition - not even light escapes them!

The first way we detect black holes is by their gravitational influence. For example, at the centre of the Milky Way, we see an empty spot where all of the stars are circling around as if they were orbiting a really dense mass. That's where the black hole is.

The second way is by observing the matter falling into the black hole. As matter falls in, it settles in a disk around the black hole that can get very hot. Some of the energy liberated from falling in is turned into light, which we can then see, for example, in X-rays.

Look Up Question 35
What is a concept called wormholes, do they really exist?

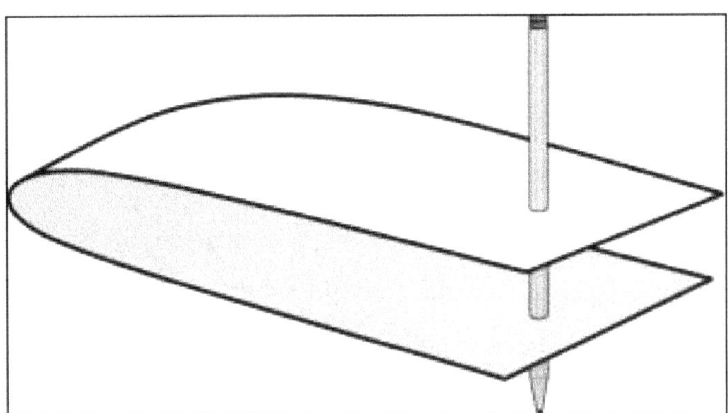

Figure: Wormhole is shortcut in the fabric of space and time. The figure describes the analogy of wormhole explained with sheet of paper.

A wormhole is a theoretical passage through space-time that could create shortcuts for long journeys across the Universe. Wormholes are predicted by the theory of general relativity. But be wary: wormholes bring with them the dangers of sudden collapse, high radiation and dangerous contact with exotic matter.

Wormhole theory

Wormholes were first theorized in 1916, though that wasn't what they were called at the time. While reviewing another physicist's solution to the equations in Albert Einstein's theory of general relativity, Austrian physicist Ludwig Flamm realised another solution was possible. He described a "white hole," a theoretical time reversal of a black hole. Entrances to both black and white holes could be connected by a space-time conduit.

In 1935, Einstein and physicist Nathan Rosen used the theory of general relativity to elaborate on the idea, proposing the existence of "bridges" through space-time. These bridges connect two different points in space-time, theoretically creating a shortcut that could reduce travel time and distance. The shortcuts came to be called Einstein-Rosen bridges, or wormholes.

Wormholes contain two mouths, with a throat connecting the two. The mouths would most likely be spheroidal. The throat might be a straight stretch, but it could also wind around, taking a longer path than a more conventional route might require.

Einstein's theory of general relativity mathematically predicts the existence of wormholes, but none have been discovered to date. A negative mass wormhole might be spotted by the way its gravity affects light that passes by.

Certain solutions of general relativity allow for the existence of wormholes where the mouth of each is a black hole. However, a naturally occurring black hole, formed by the collapse of a dying star, does not by itself create a wormhole.

Many scientists and astronomers believe that wormholes are real. Many things in astrophysics and theoretical physics have never been discovered, but, as they say, "the math works out", meaning that they are possible. It is believed that if wormholes do exist, they are the result of a primordial microscopic wormhole that was expanded at the beginning of the Universe, directly following the Big Bang.

Some people refer to them as shortcuts through the Universe, while others claim they may be a means of travelling through time. The fact is, they are fascinating and complicated and very confusing for some people, even if they have seen Interstellar.

The problem with wormholes:

So far, physicists haven't determined a way in which wormholes would form naturally in the Universe. However, theoretical physicist John Wheeler said it's possible that wormholes may spontaneously appear and disappear, according to his quantum foam hypothesis (the idea that virtual particles are, quite weirdly, popping in and out of existence at all times).

Unfortunately, Wheeler theorized that these impromptu wormholes would be super small, appearing at the Planck scale. That's about 10^{-33} centimetres long. In other words, the wormhole would be so small that it'd be almost impossible to detect.

Wormholes come with a lot of caveats, but here's an even bigger one. Wormholes could very well connect two completely different space-times; i.e. the entry point might exist in a completely different era. That means traversing via wormhole comes with the risk of winding up in a different time in the Universe's history. Some have even theorized that wormholes could connect completely different Universes altogether. Wrap your head around that one for a second.

Are we alone? What are we doing to detect life beyond Earth? Do we have life in our solar system?

Way back, an Italian physicist Enrico Fermi asked a question - given that our star, the Sun and our home, the Earth are part of a young planetary system compared to the rest of the Universe and that interstellar travel might be fairly easy to achieve. The theory says that Earth should have been visited by aliens already. Where are they? This punch-line question has become a well-known statement in the history which is recognised as Fermi's paradox.

To get to the bottom of this, consider the following fundamental facts:

- Stars are formed when stardust clouds collapse under the high gravitational force. The nuclear fusion inside the star continues throughout its lifetime. Supernova (death of a star) is a process of creation and distribution of key ingredients of life.

- The top components of Universe are Hydrogen, Helium, Carbon, Oxygen and Nitrogen and when we see what is life on Earth (like us), we will find our anatomy is also mainly built around Hydrogen, Oxygen, Carbon and Nitrogen (exception is Helium which is chemically inert).

- On average, every star in the Universe has around 2-3 planets! With at least 200 billion galaxies out there (and possibly even more), we're very likely talking about a Universe filled with around 10^{24} planets, or, for those of you who like it written out, around 1,000,000,000,000,000,000,000,000 planets in our observable Universe.

- If extremophiles can exist, why not ET? Extremophiles are organisms that have been discovered on Earth that survive in environments that were once thought not to be able to sustain life. These extreme environments include intense heat, highly acidic environments, extreme pressure and extreme cold. Different organisms have developed varying ways of adapting to these environments, but most scientists agree that it is unlikely that life on Earth originated under such extremes.

Considering above facts, do you really think we are alone in the Universe? I don't, but yes, we don't have any evidence yet to stamp it. There are many brilliant minds around the globe working on it.

Following are the efforts ongoing to find the answer of this profound and game changing question:

Microbial life in our solar system: Microbes are single-cell organisms. They are so tiny that millions can fit into the eye of a needle. They are Earth's oldest form of life, dating back more than 3.5 billion years, hundreds of millions of years before the age of dinosaurs. Could there be microbial life on other worlds in our solar system? The key efforts in finding life in our solar system are happening towards finding microbial life. There are at least seven other places in our own solar system, so kind of next door places, you could get to with a rocket, that could have microbial life.

Mars is an obvious one. It's possible that lower life forms are hidden under the dirt of the Red Planet, 30 meters (100 feet) or so below the surface where some liquid water might exist. Aside from Mars, three of Jupiter's Moons are candidates. One is Europa, which has subsurface oceans capable of hosting microbial life. These would survive off the hot spots at the bottom, which are like little mini volcanoes and that would give them energy for life. The others are, presumably, Ganymede — the largest Moon in the whole solar system and is home to a body of water similar to the Earth's oceans, but buried under a thick crust of ice — and Callisto, which has an ocean and an atmosphere. Then, there are two Moons around Saturn that could potentially harbour some form of life. One is Titan, which has liquid lakes made of natural gas. Lastly, there's Pluto. Under the surface of Pluto there could be pockets of liquid water. Any place you have liquid water, or liquid of any kind there may have microbes. These seven have the right organic processes that could serve as food or a source of energy as well as some form of liquid and not necessarily just water, to sustain microbial life. You have something that gives you food fundamentally, and the opportunity to create life, which after all is just organic chemistry. We have not yet found any microbial life outside Earth, as we have not yet visited any of these 7 places yet. But we are very close to doing so. In next one or two decades, we should have positive result from this side.

Finding Intelligent Life in the Universe

Do you know what SETI is? SETI (the Search for Extra Terrestrial Intelligence) is a scientific effort to discover intelligent life elsewhere in the Universe, primarily by attempting to discover radio signals that indicate intelligence. Cornell astronomer Frank Drake is credited with being the first to "listen" for intelligent signals with a radio telescope in 1960. The main work done by SETI is to search about 1,000 stars within 200 light-years of our solar system for radio signals beamed toward us or any other location. The 140-feet radio telescope in Green Bank, West Virginia aims at one star at a time while astronomer-monitored computers search each 1,000 band from 1,000 to 3,000 MHz for a signal limited to a narrowband range. Scientists believe that a signal focused within a narrow frequency band would suggest an intelligent source.

Apart from scanning sky for stars one by one, there are 10 more ideas that have been put forth, and even put into practice by scientists working in SETI.

1. **Optical SETI:** Russian and American scientists have been searching the skies periodically for the last couple of decades looking for laser light, which is not only distinguishable from other natural types of light, such as starlight, but could only be produced by an intelligent source.

2. **Look for huge alien structures:** When people bring this one up, the best example is always the Dyson sphere, a hypothetical structure that a civilisation would build around an entire star to capture all its energy.

3. **Find evidence of asteroid mining:** Humans are already looking at the asteroids in our solar system and considering their potential for mining, so why wouldn't an alien civilisation do the same? Evidence could include changes in the chemical composition of the asteroid, the size distribution of debris surrounding it, or other thermal changes that could be detected from Earth.

4. **Check planetary atmospheres for pollutants:** If there are non-natural chemicals, such as chlorofluorocarbons in a planet's atmosphere, it's a sign that there might be someone with technology on the ground.

5. **Look for signs of stellar engineering:** For now, this is the stuff of science fiction, but a civilization capable of tinkering with a star would surely be of interest to us Earthlings.

6. **Look for an alien artifact here on Earth:** Earth has been around for billions of years—who say that aliens haven't been here before? If they visited long ago, perhaps they left behind something in a difficult-to-reach spot, such as at the bottom of the ocean.

7. **Find a pattern in neutrinos:** Davies points out in his book that neutrinos, those ghostly subatomic particles, are probably better suited for bringing a message over a long distance than either radio or optical signals. A message would have to be simple, transmitted in a sort of alien Morse code but we could detect it here on Earth.

8. **Check for a message in DNA:** DNA is just another way to encode information. Aliens, or even just an alien probe, could have visited Earth long ago and inserted a message into some ancestral creature. Of course, there are several hurdles to such an idea, as Davies notes, getting the message here, getting it into a critter, keeping it from getting destroyed by mutations over perhaps millions of years. But it certainly is an intriguing possibility.

9. **Find a propulsion signature from an alien spacecraft:** Hey, if it worked for the Vulcans in Star Trek, why not us?

10. **Invite ET to log on:** A group of scientists have set up a web site asking for an extra-terrestrial intelligence to send them an e-mail. So far, all the responses have been hoaxes, but asking for a shout-out never really hurts.

Let's now consider probability of ET life with famous Drake Equation's perspective. But first what is Drake Equation?

The Drake Equation is an attempt to encapsulate all the variables that would be relevant to establishing the number of intelligent civilisations that existed in the Milky Way galaxy and which are broadcasting radio signals at this particular point in time. The Drake Equation is composed of seven terms. The first six terms are used to compute the rate at which intelligent civilisations are being created and the final term identifies how long each one lasts on an average as a broadcasting civilisation. It is worth stressing that the Drake Equation applies only to intelligent civilisations in the Milky Way galaxy. It does not apply to civilizations in other galaxies because they are too distant to be able to detect their radio signals.

The Drake Equation is:

Figure: Famous Drake equation

Where:

N = The number of broadcasting civilizations.

R = Average rate of formation of suitable stars (stars/year) in the Milky Way galaxy

f_p= Fraction of stars that form planets

n_e = Average number of habitable planets per star

f_l = Fraction of habitable planets (ne) where life emerges

f_i = Fraction of habitable planets with life where intelligent evolves

f_c = Fraction of planets with intelligent life capable of interstellar communication

L = Years a civilisation remains detectable

There are seven variables to this equation, and the fact is we still don't know best possible numbers for each one of them. We can do some approximate guess for each and find out odds we have. But the fact that Universe real estate is so big, we for sure can confidently say that we should have thousands (if not million) of intelligent civilisation in our Galaxy itself.

Look Up Question 37
Do other stars also have their planetary system? How did we discover exoplanets?

For many years scientists have studied our own solar system. But until the last few years, we knew of no other solar systems.

This fact may seem surprising as the Sun is one of about 200 billion stars (or perhaps more) just in the Milky Way galaxy alone. With all those other stars, why haven't scientists studied other solar systems, at least enough to know how many are in our galaxy?

Well, the reason is that planets around other stars are really hard to find. Planets shine only by the light they reflect from the star they orbit, and they don't reflect much light. And the stars, along with any planets under their control, are so far away that picking out a faint planet near a distant star is like spotting a mosquito next to a big searchlight mile away.

Finally, in the middle 1990s, astronomers found strong evidence of planets around other stars. In all cases, they found the planets not by taking pictures of them, but rather by detecting their astonishingly gentle tugs on the stars they orbit. Although the star holds the planet tightly in its gravitational grip, the planet also exerts a gravitational pull back on the star, and that is what astronomers' measure. It amounts to seeing the star wobble back and forth very slightly as the planet completes each orbit. Learn more about this gravitational dance as you try to solve the extraterrestrial riddle.

Figure: NASA's Kepler telescope.

After that astronomers started detecting planets through several other methods as well. For example, if the orbit of a planet happens to be aligned so that planet occasionally travels in front of the star from our perspective on Earth, it blocks some of the light. Even though the planet is tiny compared to the star, extremely sensitive instruments can measure the tiny change in brightness. NASA's Kepler mission used this technique to identify hundreds of stars that may have planets. Astronomers are observing these stars more carefully to confirm the presence of the candidate planets.

NASA is working on more space missions that will allow scientists not only to find other solar systems but also to study the planets there in greater details. Some of the intriguing questions these missions might help answer are how common are other solar systems; is our solar system typical, with giant planets like Jupiter and smaller ones like Earth; how does solar systems form and evolve; are there other planets capable of supporting life; and is there life on other planets?

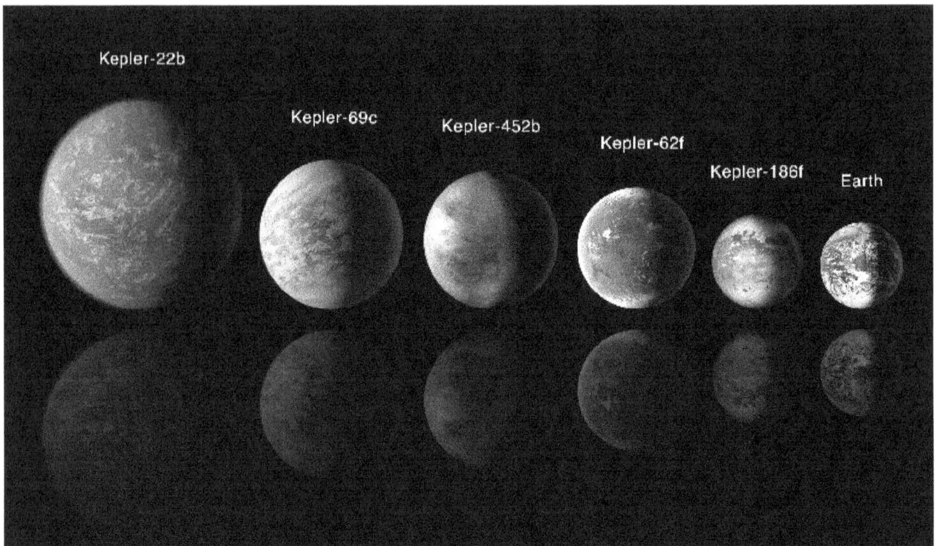

Figure: Some of the exoplanets discovered by Kepler in past few years.

So far, astronomers have found more than 500 solar systems and are discovering new ones every year. Given how many they have found in our own neighbourhood of the Milky Way galaxy, scientists estimate that there may be tens of billions of solar systems in our galaxy, perhaps even as many as 100 billion.

Whether this estimate is correct and how similar other solar systems are to ours, remains to be seen. It has only been a few years since the first solar system apart from ours was detected, and they are still extremely difficult to study, so this whole subject is still in its infancy. By the time our friends who asked the questions are adults, we will know a great deal more.

Perhaps someday you will help find the answers. And even if you don't, you may grow up in a time when humankind has a much clearer idea of how we and our home planet fit into the Cosmos.

In the early 1990s, one thing was certain about the expansion of the Universe. It might have enough energy density to stop its expansion and re-collapse, it might have so little energy density that it would never stop expanding, but gravity was certain to slow the expansion as time went on. Granted, the slowing had not been observed, but, theoretically, the Universe had to slow. The Universe is full of matter and the attractive force of gravity pulls all matter together. Then came 1998 and the Hubble Space Telescope (HST) observations of very distant supernovae that showed that, a long time ago, the Universe was actually expanding more slowly than it is today. So, the expansion of the Universe has not been slowing due to gravity, as everyone thought, it has been accelerating. No one expected this; no one knew how to explain it. But something was causing it.

Eventually theorists came up with three explanations. Maybe it was a result of a long-discarded version of Einstein's theory of gravity, one that contained what was called a "cosmological constant." Maybe there was some strange kind of energy-fluid that filled space. Maybe there is something wrong with Einstein's theory of gravity and a new theory could include some kind of field that creates this cosmic acceleration. Theorists still don't know what the correct explanation is, but they have given the solution a name. It is called dark energy.

What Is Dark Energy?

More is unknown than is known. We know how much dark energy there is because we know how it affects the Universe's expansion. Other than that, it is a complete mystery. But it is an important mystery. It turns out that roughly 73% of the Universe is dark energy. Dark matter makes up about 23%. The rest, everything on Earth, everything ever observed with all of our instruments, all normal matter, adds up to less than 4% of the Universe. Come to think of it, maybe it shouldn't be called "normal" matter at all, since it is such a small fraction of the Universe.

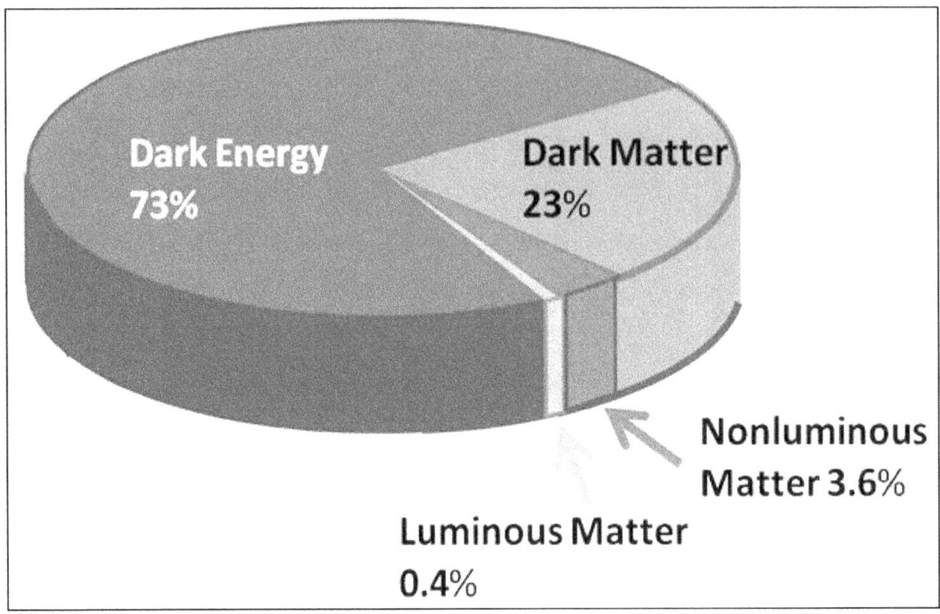

Figure: Universe is made of Dark Energy and Dark Matter; we all are just part of 4%!

One explanation for dark energy is that it is a property of space. Albert Einstein was the first person to realise that empty space is not anything. Space has amazing properties, many of which are just beginning to be understood. The first property that Einstein discovered is that it is possible for more space to come into existence. Then one version of Einstein's gravity theory, the version that contains a cosmological constant, makes a second prediction: "empty space" can possess its own energy. Because this energy is a property of space itself, it would not be diluted as space expands. As more space comes into existence, more of this energy-of-space would appear. As a result, this form of energy would cause the Universe to expand faster and faster. Unfortunately, no one understands why the cosmological constant should even be there, much less why it would have exactly the right value to cause the observed acceleration of the Universe.

Another explanation for how space acquires energy comes from the quantum theory of matter. In this theory, "empty space" is full of temporary ("virtual") particles that continually form and then disappear. But when physicists tried to calculate how much energy this would give empty space; the answer came out wrong - wrong by a lot. The number came out 10^{120} times too big. That's a 1 with 120 zeros after it. It's hard to get an answer that bad. So, the mystery continues.

Another explanation for dark energy is that it is a new kind of dynamical energy fluid or field, something that fills all of space but something whose effect on the expansion of the Universe is the opposite of that of matter and normal energy. Some theorists have named this "quintessence," after the fifth element of the Greek philosophers. But, if quintessence is the answer, we still don't know what it is like, what it interacts with, or why it exists. So, the mystery continues.

A last possibility is that Einstein's theory of gravity is not correct. That would not only affect the expansion of the Universe, but it would also affect the way that normal matter in galaxies and clusters of galaxies behaved. This fact would provide a way to decide if the solution to the dark energy problem is a new gravity theory or not: we could observe how galaxies come together in clusters. But if it does turn out that a new theory of gravity is needed, what kind of theory would it be? How could it correctly describe the motion of the bodies in the Solar System, as Einstein's theory is known to do, and still give us the different prediction for the Universe that we need? There are candidate theories, but none are compelling. So, the mystery continues.

The thing that is needed to decide between dark energy possibilities - a property of space, a new dynamic fluid, or a new theory of gravity - is more data, better data.

What Is Dark Matter?

By fitting a theoretical model of the composition of the Universe to the combined set of cosmological observations, scientists have come up with the composition that we described above, ~73% dark energy, ~23% dark matter, ~4% normal matter. What is dark matter?

We are much more certain what dark matter is not than we are about what it is. First, it is dark, meaning that it is not in the form of stars and planets that we see. Observations show that there is far too little visible matter in the Universe to make up the 23% required by the observations. Second, it is not in the form of dark clouds of normal matter, matter made up of particles called baryons. We know this because we would be able to detect baryonic clouds by their absorption of radiation passing through them. Third, dark matter is not antimatter, because we do not see the unique gamma rays that are produced when antimatter annihilates with matter. Finally, we can rule out large galaxy-sized black holes on the basis of how many gravitational lenses we see. High concentrations of matter bend light passing near them from objects further away, but we do not see enough lensing events to suggest that such objects to make up the required 23% dark matter contribution.

However, at this point, there are still a few dark matter possibilities that are viable. Baryonic matter could still make up the dark matter if it were all tied up in brown dwarfs or in small, dense chunks of heavy elements. These possibilities are known as massive compact halo objects, or "MACHOs". But the most common view is that dark matter is not baryonic at all, but that it is made up of other, more exotic particles like axions or WIMPS.

Look Up Question 39
What is LHC and how it's important for new discoveries in astrophysics. What is God particle?

The Large Hadron Collider (LHC) is a particle accelerator developed by CERN, the world's largest organisation devoted to particle physics. A particle accelerator, sometimes called an "atom smasher" by lay people, is a device that propels subatomic particles called hadrons at high speeds. Machines such as the LHC make it possible to split particles into smaller and smaller components in the quest for the identification of so-called elementary particles, from which all matter and energy might derive.

Figure: Location of LHC at CERN

The LHC, located at CERN headquarters, became operational in 2008. In operation, the LHC replicates, on a miniature scale, the conditions existing in the Universe a tiny fraction after the Big Bang. Thus, it may be possible to discern what happened in the early evolutional stages of the Universe. Among other things (though not yet seen), the LHC may yield evidence of further dimensions beyond our familiar four (three spatial dimensions, plus time).

The LHC is expected to help physicists, astronomers and cosmologists answer questions about the nature and origins of matter, energy and the Universe. For example:

- Is antimatter simply a "mirror image" of matter or is the relationship more complex?
- Why does matter seem to predominate over antimatter in the Universe?
- Why didn't all the matter and antimatter combine long ago, converting the whole Universe into energy?
- What is the nature of dark matter?

- Why do only some particles have mass?

Figure: LHC tube

The LHC uses intense magnetic fields generated by superconductivity to accelerate hadrons in circular path 27 kilometres (about 17 miles) in circumference. The particles interact with the magnetic fields to gain energy with each revolution. The LHC is capable of accelerating protons to energy levels of about 14 TeV (trillion electronvolts, where a trillion is equal to 10^{12}) or 2.2×10^{-6} joule. Nuclei of lead atoms are accelerated to speeds sufficient to cause collisions having energy levels near 1150 TeV or 1.8×10^{-4} joule. The electronvolt (eV) is the amount of kinetic energy gained by an electron passing through an electrostatic field producing a potential difference of one volt. The joule is the standard unit of energy, equivalent to a watt - second.

Figure: LHC cross sectional view

The term 'The God particle' was coined by the physicist Leon Lederman in his 1993 popular science book, The God Particle: If the Universe Is the Answer, What Is the Question? The particle that the book title refers to is the 'Higgs boson'.

What is the Higgs boson?

The theories and discoveries of thousands of physicists over the past century have resulted in a remarkable insight into the fundamental structure of matter: everything in the Universe is found to be made from twelve basic building blocks called fundamental particles, governed by four fundamental forces.

Our best understanding of how these twelve particles and three of the forces are related to each other is encapsulated in the Standard Model of particles and forces. Developed in the 1960s and 70s, it has successfully explained a host of experimental results and precisely predicted a wide variety of phenomena. Over time and through many experiments by many physicists, the Standard Model has become established as a well-tested physics theory.

In the 1960s physicists began to realise that there are very close ties between two of the four fundamental forces between elements: the weak force and the electromagnetic force. The two forces could be described within a unified theory, which forms the basis of the Standard Model. This unification implies that electricity, magnetism, light and some types of radioactivity are all manifestations of a single underlying force called the electroweak force.

However, in order for this unification to work mathematically, it requires that force-carrying particles have no mass. We know from experiments that this is not true. Physicists including Peter Higgs, Robert Brout and François Englert came up with a solution to solve this conundrum.

They suggested that all particles had no mass just after the Big Bang. As the Universe cooled and the temperature fell below a critical value, an invisible force field which became commonly known as the 'Higgs field' was formed together with the associated particle, now called the 'Higgs boson'. The field prevails throughout the Cosmos: any particles that interact with it are given a mass via the Higgs boson. The more they interact, the heavier they become, whereas particles that never interact are left with no mass at all.

This idea provided a satisfactory solution and fitted well with established theories and phenomena. The problem is that no one has ever observed the Higgs boson in an experiment to confirm the theory. Finding this particle would give an insight into why certain particles have mass, and help to develop subsequent physics.

However, we do not know the mass of the Higgs boson itself, which makes it more difficult to identify. Physicists have to look for it by systematically searching a range of mass within which it is predicted to exist. The yet unexplored range is accessible using the Large Hadron Collider, which will determine the existence of the Higgs boson. If it turns out that we cannot find it, something else will appear that does the same job. Either way, our knowledge of the Universe will advance.

What does the Higgs boson have to do with God?

The Higgs boson has nothing to do with God. It was simply a snappy term to illustrate the ubiquitous effect of the Higgs field, and its importance in determining mass.

Look Up Question 40
What is LIGO and how it has detected collision of black holes and neutron stars?

LIGO stands for Laser Interferometer Gravitational-wave Observatory. It is the world's largest physics experiment laboratory that assesses the physical characteristics of light to detect and study the origin of gravitational waves.

Figure: LIGO observatory location

Though it is described as a gravitational wave observatory, it differs from other observatories in the world. It is not round in shape and does not have a telescopic dome unlike in other observatories. It does not follow electromagnetic radiation by employing optical or radio telescopes. Laser Interferometer Gravitational-Wave Observatory employs two gigantic laser interferometers located thousands of kilometres apart to study gravitational waves. Despite the word observatory in name, it is one of the highly complicated physics laboratories in the world housing giant particle accelerators. For LIGO, focus is only on invisible gravitational waves.

LIGO is a collaborative project that consists of nine thousand researchers from twenty countries. It is funded by U.S National Science Foundation. Australia, Germany and United Kingdom also support financially. California Institute of Technology (Caltech) and the Massachusetts Institute of Technology (MIT) control operations. LIGO was established with the mission of opening the field of gravitational-wave astrophysics through identification of the waves.

In LIGO, Laser interferometry is used to observe distortions occurring between stationary, hanging masses because of passage of gravitational waves. First gravitational wave detectors became operational in 1999. During the first phase between 2002 and 2010, they could not identify any gravitational waves. However, this experience provided inputs for further research in next phase. During this phase, the instruments were redesigned and interferometers became 10 times more sensitive. This made the equipment to listen 10 time farther away waves, giving access to 1000(10x10x10) times more volume of space. The deeper search for gravitational waves with the revised interferometers started in September 2015 and within days, on September 14, 2015, the LIGO detectors succeeded in the first detection of gravitational waves that too originated around 1.3 billion light years back from the collusion of two black holes.

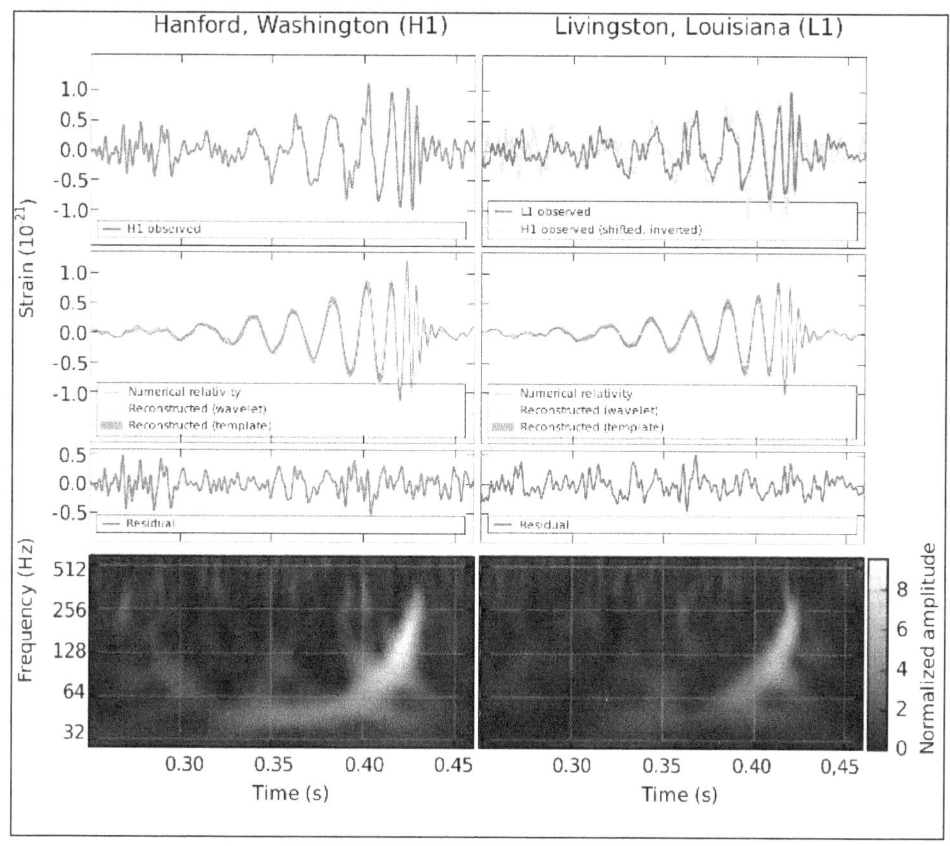

Figure: 2 Black Holes collision signal detected by LIGO which fetched Nobel prize.

The hard work, dedication and continuous efforts of the researchers finally led to the identification of gravitational waves. The 2017 physics Nobel Laureates contributed immensely for the success of the project.

LIGO-India is a planned advanced gravitational-wave observatory to be located in India as part of the worldwide network. The project recently received the in-principle approval from the Indian government. LIGO-India is planned as a collaborative project between a consortium of Indian research institutions and the LIGO Laboratory in the USA, along with its international partners Australia, Germany and the UK.

The discovery: Gravitational waves from neutron stars

On Aug. 17, advanced LIGO and advanced Virgo (the current upgraded versions of both observatories) detected a gravitational-wave signal possessing an extraordinary amount of energy — "something like a billion times the energy of the luminosity of the Milky Way. "Its energy was enough to outshine the 100 billion stars in our galaxy by about a billion-fold for the 50 or so seconds it took place. This event is the first-time scientists have witnessed two neutron stars merging. One main clue that the signal came from such a merger was its duration, the longest gravitational-wave signal detected to date.

Black holes are denser than neutron stars, so the signals from their mergers are relatively brief. "Previously detected black-hole mergers lasted for a second, maybe two seconds.

So far, LIGO has detected four black-hole mergers and one neutron-star merger. Some researchers had predicted neutron-star mergers would be more common than black-hole mergers, whereas others had predicted the opposite. While neutron-star mergers are more common in any given volume, black-hole mergers are more energetic "and so can get detected from much farther out."

Look Up Question 41
Do we have theory of everything (string theory) which combines the theory of big (classical mechanics) and the theory of small (quantum mechanics)?
How many dimensions may we have per the string theory?

String theory, in particle physics is a theory that attempts to merge quantum mechanics with Albert Einstein's general theory of relativity. The name string theory comes from the modelling of subatomic particles as tiny one-dimensional "string-like" entities rather than the more conventional approach in which they are modelled as zero-dimensional point particles. The theory envisions that a string undergoing a particular mode of vibration corresponds to a particle with definite properties such as mass and charge. In the 1980s, physicists realized that string theory had the potential to incorporate all four of nature's forces - gravity, electromagnetism, strong force, and weak force, and all types of matter in a single quantum mechanical framework, suggesting that it might be the long-sought unified field theory. While string theory is still a vibrant area of research that is undergoing rapid development, it remains primarily a mathematical construct because it has yet to make mark with experimental observations.

Relativity and Quantum Mechanics

In 1905, Einstein unified space and time with his special theory of relativity, showing that motion through space affects the passage of time. In 1915, Einstein further unified space, time, and gravitation with his general theory of relativity, showing that warps and curves in space and time are responsible for the force of gravity. These were monumental achievements, but Einstein dreamed of an even grander unification. He envisioned one powerful framework that would account for space, time, and all of nature's forces—something he called a unified theory. For the last three decades of his life, Einstein relentlessly pursued this vision. Although from time to time rumours spread that he had succeeded, closer scrutiny always dashed such hopes. Most of Einstein's contemporaries considered the search for a unified theory to be a hopeless, if not misguided, quest.

In contrast, the primary concern of theoretical physicists from the 1920s onward was quantum mechanics—the emerging framework for describing atomic and subatomic processes. Particles at these scales have such tiny masses that gravity is essentially irrelevant in their interactions, and so for decades, quantum mechanical calculations generally ignored general relativistic effects. Instead, by the late 1960s the focus was on a different force, the strong force, which binds together the protons and neutrons within the atomic nuclei. Gabriele Veneziano, a young theorist working at the European Organization for Nuclear Research (CERN), contributed a key breakthrough in 1968 with his realization that a 200-year-old formula, the Euler beta function, was capable of explaining much of the data on the strong force then being collected at various particle accelerators around the world. A few years later, three physicists— Leonard Susskind of Stanford University, Holger Nielsen of the Niels Bohr Institute, and Yoichiro Nambu of the University of Chicago, significantly amplified Veneziano's insight by showing that the mathematics underlying his proposal described the vibrational motion of minuscule filaments of energy that resemble tiny strands of string, inspiring the name string theory. Roughly speaking, the theory suggested that the strong force amounted to strings tethering together particles attached to the strings' endpoints.

Dimensions and Vibrations

In 1984, string theorists achieved a major breakthrough. Through a remarkable calculation, they proved that the equations of string theory were consistent after all. By the time word of this result had spread throughout the physics community, hundreds of researchers had dropped what they were working on and turned their full attention to the string theory.

Within a few months, string theory's unified framework took shape. Much as different vibrational patterns of a violin string play different musical notes, the different vibrations of the tiny strands in the string theory were imagined yielding different particles of nature. Per the theory, the strings are so small that they appear to be points, as particles had long been thought to be, but in reality, they have length (about 10^{-33}cm); the mass and the charge of a particle is determined by how a string vibrates. For example, string theory posits that an electron is a string undergoing one particular vibrational pattern; a quark is imagined as a string undergoing a different vibrational pattern. Crucially, among the vibrational patterns, physicists argued, would also be the particles found by experiment to communicate nature's forces. Thus, string theory was proposed as the sought-for unification of all forces and all matter.

What about the six extra spatial dimensions required by string theory? Following a suggestion made in the 1920s by Theodor Kaluza of Germany and

Oskar Klein of Sweden, string theorists envisioned that dimensions come in two distinct varieties. Like the unfurled length of a long garden hose, dimensions can be big and easy to see. But like the shorter, circular girth of the garden hose, dimensions can also be far smaller and more difficult to detect. This becomes more apparent by imagining that the circular cross section of the garden hose is shrunk ever smaller, below what can be seen with the naked eye, misleading a casual observer into thinking the garden hose has only one dimension, its length. Similarly, according to string theory, the three dimensions of common experience are large and manifest, while the other six dimensions are crumpled so small that they have so far evaded detection.

During the decade from 1984 to 1994, many theoretical physicists strove to fulfil string theory's promise by developing this abstract, wholly mathematical framework into a concrete, predictive theory of nature. Because the infinitesimal size of strings has precluded their direct detection, theorists have sought to extract indirect implications of the theory that might be testable. In this regard, the extra dimensions of string theory have proved a major hurdle. Imagining these extra dimensions as small and hidden is a reasonable explanation for their apparent absence. Nevertheless, their detailed geometry is required for the theory to offer predictions. The reason is that strings are so small that they would vibrate within the tiny extra dimensions. Studies showed that, much as the shape and size of a French horn affect the vibrational patterns of airstreams coursing through the instrument, the exact shape and size of the extra dimensions would affect how strings vibrate. And since the strings' vibrations determine quantities such as particle masses and charges, productivity requires knowledge of the geometric form of the extra dimensions. Unfortunately, the equations of string theory allow the extra dimensions to take many different geometric forms, making it difficult to extract definitive testable predictions.

M-Theory and AdS/CFT Correspondence

By the mid-1990s these and other obstacles were again eroding the ranks of string theorists. But in 1995 another breakthrough reinvigorated the field.

Edward Witten of the Institute for Advanced Study, building on contributions of many other physicists, proposed a new set of techniques that refined the approximate equations on which all the work in string theory had been based. These techniques helped reveal a number of new features of string theory, including the realization that the theory has not six but seven extra spatial dimensions. The more exact equations also revealed ingredients in string theory besides strings, membrane-like objects of various dimensions, collectively called branes. Finally, the new techniques established that various versions of string theory developed over the preceding decades were essentially all the same. Theorists call this unification of formerly distinct string theories by a new name, M-theory, with the meaning of M being deferred until the theory is more fully understood.

Another advance in string theory happened in 1997 when Juan Maldacena of Harvard University discovered the anti-de Sitter/conformal field theory (AdS/CFT) correspondence. Maldacena found that a string theory operating with a particular environment (involving a space-time known as an anti-de Sitter space) was equivalent to a type of quantum field theory operating in an environment with one less spatial dimension. This has proved to be one of the most profound discoveries in string theory, establishing a powerful link to the more conventional methods of quantum field theory, providing an exact mathematical formulation of string theory in certain environments, and inspiring thousands of further technical studies.

Today the understanding of many facets of string theory is still in its formative stage. Researchers recognize that, although remarkable progress has been made over the past five decades, collectively the work is burdened by its piecemeal development, with incremental discoveries having been joined like pieces of a jigsaw puzzle. That the pieces fit coherently is impressive, but the larger picture they are filling out, the fundamental principle underlying the theory still remains mysterious. Equally pressing, the theory has yet to be supported by observations and hence remains a totally theoretical construct.

Supersymmetry and Cosmological Signature

One essential quality of string theory is known as supersymmetry, a mathematical property that requires every known particle species to have a partner particle species, called a superpartner. (This property accounts for string theory, often being referred to as, superstring theory.) As yet, no superpartner particles have been detected experimentally, but researchers believe this may be due to their weight: they are heavier than their known counterparts and require a machine at least as powerful as the Large Hadron Collider at CERN to produce them. If the superpartner particles are found, it is still not sufficient to prove the string theory correct, because more-conventional point-particle theories have also successfully incorporated supersymmetry into their mathematical structure. However, the discovery of supersymmetry would confirm an essential element of the string theory and give circumstantial evidence that this approach to unification is on the right track.

Even if these accelerator-based tests are inconclusive, there is another way that string theory may one day be tested. Through its impact on the earliest, most extreme moments of the Universe, the physics of string theory may have left faint cosmological signatures, for example, in the form of gravitational waves or a particular pattern of temperature variations in the cosmic microwave background radiation that may be observable by the next generation of precision satellite-borne telescopes and detectors. It would be a fitting conclusion to Einstein's quest for unification if a theory of the smallest microscopic component of matter were confirmed through observations of the largest astronomical realms of the Cosmos.

Look Up Question 42
What is singularity?

Ever since scientists first discovered the existence of black holes in our Universe, we have all wondered: what could possibly exist beyond the veil of that terrible void? In addition, ever since the theory of General Relativity was first proposed, scientists have been forced to wonder, what could have existed before the birth of the Universe – i.e. before the Big Bang?

Interestingly enough, these two questions have come to be resolved (after a fashion) with the theoretical existence of something known as a Gravitational Singularity – a point in space-time where the laws of physics, as we know them, break down. And while there remain challenges and unresolved issues about this theory, many scientists believe that beneath veil of an event horizon, and at the beginning of the Universe, this was what existed.

Definition

In scientific terms, a gravitational singularity (or space-time singularity) is a location where the quantities that are used to measure the gravitational field become infinite in a way that does not depend on the coordinate system. In other words, it is a point in which all the physical laws are indistinguishable from one another, where space and time are no longer interrelated realities, but merge indistinguishably and cease to have any independent meaning.

Origin of Theory

Singularities were first predicated as a result of Einstein's Theory of General Relativity, which resulted in the theoretical existence of black holes. In essence, the theory predicted that any star reaching beyond a certain point in its mass would exert a gravitational force so intense that it would collapse.

At this point, nothing would be capable of escaping its surface, including light. This is due to the fact that the gravitational force would exceed the speed of light in vacuum i.e. 299,792,458 meters per second (1,079,252,848.8 km/h; 670,616,629 mph). This phenomenon is known as the Chandrasekhar Limit, named after the Indian astrophysicist Subrahmanyan Chandrasekhar, who proposed it in 1930. At present, the accepted value of this limit is believed to be 1.39 Solar Masses (i.e. 1.39 times the mass of our Sun), which works out to a whopping 2.765×10^{30} kg (or 2,765 trillion trillion metric tons).

Another aspect of modern General Relativity is that at the time of the Big Bang, the Universe was a singularity. Roger Penrose and Stephen Hawking both developed theories that attempted to answer how gravitation could produce singularities, that eventually merged together to be known as the Penrose–Hawking Singularity Theorems.

Per the Penrose Singularity Theorem that proposed in 1965, a time-like singularity will occur within a black hole whenever matter reaches certain energy conditions. At this point, the curvature of space-time within the black hole becomes infinite, thus turning it into a trapped surface where time ceases to function.

The Hawking Singularity Theorem added to this by stating that a space-like singularity can occur when matter is forcibly compressed to a point, causing the rules that govern matter to break down. Hawking traced this back in time to the Big Bang, which he claimed was a point of infinite density. However, Hawking later revised this to claim that general relativity breaks down at times prior to the Big Bang, and hence no singularity could be predicted by it.

Some more recent proposals also suggest that the Universe did not begin as a singularity. These include theories like Loop Quantum Gravity, which attempts to unify the laws of quantum physics with gravity. This theory states that, due to quantum gravity effects, there is a minimum distance beyond which gravity no longer continues to increase, or that interpenetrating particle waves mask gravitational effects that would be felt at a distance.

Types of Singularities:

The two most important types of space-time singularities are known as Curvature Singularities and Conical Singularities. Singularities can also be divided per whether they are covered by an event horizon or not. In the case of the former, you have the Curvature and Conical; whereas in the latter, you have what are known as Naked Singularities.

A Curvature Singularity is best exemplified by a black hole. At the centre of a black hole, space-time becomes a one-dimensional point which contains a huge mass. As a result, gravity become infinite and space-time curves infinitely, and the laws of physics as we know them, cease to function.

Conical singularities occur when there is a point where the limit of every general covariance quantity is finite. In this case, space-time looks like a cone around this point, where the singularity is located at the tip of the cone. An example of such a conical singularity is a cosmic string, a type of hypothetical one-dimensional point that is believed to have formed during the early Universe.

And, as mentioned, there is the Naked Singularity, a type of singularity which is not hidden behind an event horizon. These were first discovered in 1991 by Shapiro and Teukolsky using computer simulations of a rotating plane of dust that indicated that General Relativity might allow for "naked" singularities.

In this case, what actually transpires within a black hole (i.e. its singularity) would be visible. Such a singularity would theoretically be what existed prior to the Big Bang. The key word here is theoretical, as it remains a mystery what these objects would look like.

For the moment, singularities and what actually lies beneath the veil of a black hole remains a mystery. As time goes on, it is a hope that astronomers will be able to study black holes in greater detail. It is also hoped that in the coming decades, scientists will find a way to merge the principles of quantum mechanics with gravity, and that this will shed further light on how this mysterious force operates.

Look Up Question 43
Is there only one Universe or we live in multiverse?
What are theories behind multiverse?

The Universe we live in may not be the only one out there. In fact, our Universe could be just one of an infinite number of Universes making up a "multiverse."

Though the concept may stretch credulity, there's good physics behind it. And there's not just one way to get to a multiverse. Numerous physics theories independently point to such a conclusion. In fact, some experts think the existence of hidden Universes is more likely than not.

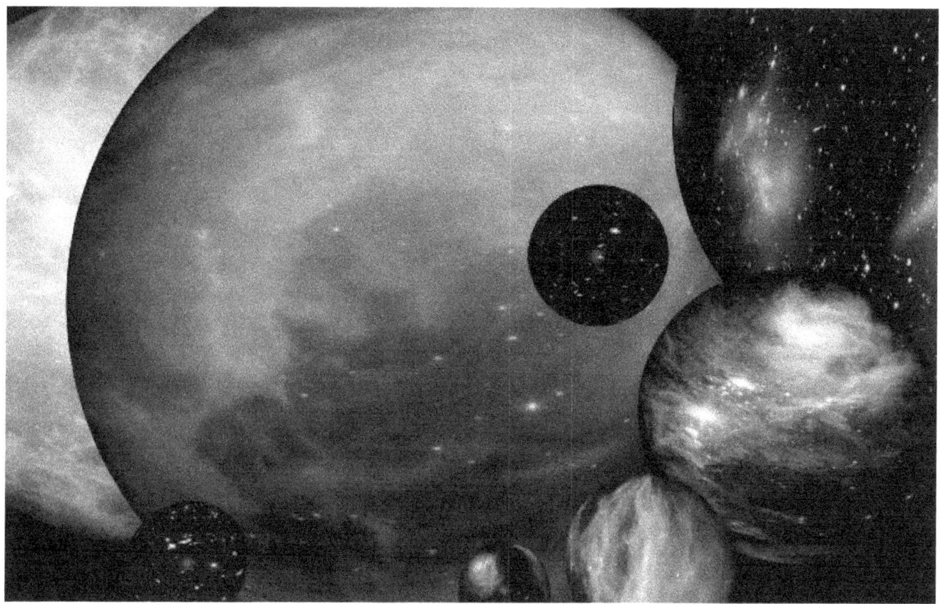

Figure: Artist description of multiverse.

Here are the five most plausible scientific theories suggesting we live in a multiverse:

1. Infinite Universes

Scientists can't be sure what the shape of space-time is, but most likely, it's flat (as opposed to spherical or even donut-shape) and stretches out infinitely. But if space-time goes on forever, then it must start repeating at some point, because there are a finite number of ways particles can be arranged in space and time.

So, if you look far enough, you would encounter another version of you — in fact, infinite versions of you. Some of these twins will be doing exactly what you're doing right now, while others will have worn a different sweater this morning, and still others will have made vastly different career and life choices.

As the observable Universe extends only as far as light has had a chance to get in the 13.7 billion years, since the Big Bang (that would be 13.7 billion light-years), the space-time beyond that distance can be considered to be its own separate Universe. In this way, a multitude of Universes exists next to each other in a giant patchwork quilt of Universes.

2. Bubble Universes

In addition to the multiple Universes created by infinitely extending space-time, other Universes could arise from a theory called "eternal inflation." Inflation is the notion that the Universe expanded rapidly after the Big Bang, in effect inflating like a balloon. Eternal inflation, first proposed by Tufts University cosmologist Alexander Vilenkin, suggests that some pockets of space stop inflating, while other regions continue to inflate, thus giving rise to many isolated "bubble Universes."

Thus, our own Universe, where inflation has ended, allowing stars and galaxies to form, is but a small bubble in a vast sea of space, some of which is still inflating, that contains many other bubbles like ours. And in some of these bubble Universes, the laws of physics and fundamental constants might be different than in ours, making some Universes strange places indeed.

3. Parallel Universes

Another idea that arises from string theory is the notion of "braneworlds", parallel Universes that hover just out of reach of our own, proposed by Princeton University's Paul Steinhardt and Neil Turok of the Perimeter Institute for Theoretical Physics in Ontario, Canada. The idea comes from the possibility of many more dimensions to our world than the three of space and one of time that we know. In addition to our own three-dimensional "brane" of space, other three-dimensional branes may float in a higher-dimensional space.

Columbia University physicist Brian Greene describes the idea as the notion that "our Universe is one of the potentially numerous 'slabs' floating in a higher-dimensional space, much like a slice of bread within a grander cosmic loaf," in his book "The Hidden Reality" (Vintage Books, 2011).

A further wrinkle on this theory suggests these brane Universes aren't always parallel and out of reach. Sometimes, they might slam into each other, causing repeated Big Bangs that reset the Universes over and over again.

4. Daughter Universes

The theory of quantum mechanics, which reigns over the tiny world of subatomic particles, suggests another way multiple Universes might arise. Quantum mechanics describes the world in terms of probabilities, rather than definite outcomes. And the mathematics of this theory might suggest that all possible outcomes of a situation do occur in their own separate Universes. For example, if you reach a crossroads where you can go right or left, the present Universe gives rise to two daughter Universes: one in which you go right, and one in which you go left.

"And in each Universe, there's a copy of you witnessing one or the other outcome, thinking — incorrectly — that your reality is the only reality," Greene wrote in his book, "The Hidden Reality."

5. Mathematical Universes

Scientists have debated whether mathematics is simply a useful tool for describing the Universe, or whether math itself is the fundamental reality, and our observations of the Universe are just imperfect perceptions of its true mathematical nature. If the latter is the case, then perhaps the particular mathematical structure that makes up our Universe isn't the only option, and in fact all possible mathematical structures exist as their own separate Universes.

"A mathematical structure is something that you can describe in a way that's completely independent of human baggage," said Max Tegmark of MIT, who proposed this brain-twisting idea. "I really believe that there is this Universe out there that can exist independently of me that would continue to exist even if there were no humans."

<h1 style="text-align:center">Look Up Question 44</h1>
<h2 style="text-align:center">What is time travel? Is it possible?</h2>

Time travel, moving between different points in time has been a popular topic for science fiction for decades. Media series from "Doctor Who" to "Star Trek" to "Back to the Future" have seen humans get in a vehicle of some sort and arrive in the past or future, ready to take on new adventures.

The reality, however, is more muddled. Not all scientists believe that time travel is possible. Some even say that an attempt would be fatal to any human who chooses to undertake it.

What is time? While most people think of time as a constant, physicist Albert Einstein showed that time is an illusion; it is relative. It can vary for different observers depending on their speed through space. To Einstein, time is the "fourth dimension." Space is described as a three-dimensional arena, which provides a traveller with coordinates such as length, width and height, showing the location. Time provides another coordinate direction although conventionally, it only moves forward. Most physicists think that time is a subjective illusion, but what if time is real?

Einstein's theory of special relativity says that time slows down or speeds up depending on how fast you move relative to something else. Approaching the speed of light, a person inside a spaceship would age much slower than his twin at home. Also, under Einstein's theory of general relativity, gravity can bend time.

Through the wormhole

General relativity also provides scenarios that could allow travellers to go back in time, according to NASA. The equations, however, might be difficult to physically achieve.

One possibility could be to go faster than light, which travels at 186,282 miles per second (299,792 kilometres per second) in a vacuum. Einstein's equations, though, show that an object at the speed of light would have both infinite mass and a length of 0. This appears to be physically impossible, although some scientists have extended his equations and said it might be done.

A linked possibility would be to create "wormholes" between points in space-time. While Einstein's equations provide for them, they would collapse very quickly and would only be suitable for very small particles. Also, scientists haven't observed these wormholes yet. Also, the technology needed to create a wormhole is far beyond anything we have today.

Alternate time travel theories

While Einstein's theories appear to make time travel difficult, some groups have proposed alternate solutions to jump back and forth in time.

Infinite cylinder

Astronomer Frank Tipler proposed a mechanism (sometimes known as a Tipler Cylinder) where one would take matter that is 10 times the Sun's mass, then roll it into very long but very dense cylinder.After spinning this up a few billion revolutions per minute, a spaceship nearby, following a very precise spiral around this cylinder, could get itself on a "closed, time-like curve". There are limitations with this method, however, including the fact that the cylinder needs to be infinitely long for this to work.

Black holes

Another possibility would be to move a ship rapidly around a black hole, or to artificially create that condition with a huge, rotating structure.

"Around and around they'd go, experiencing just half the time of everyone far away from the black hole. The ship and its crew would be travelling through time," physicist Stephen Hawking wrote in the Daily Mail in 2010.

Imagine they circled the black hole for five of their years. Ten years would pass elsewhere. When they got home, everyone on Earth would have aged five years more than they had.

Cosmic strings

Another theory for potential time travellers involves something called cosmic strings, narrow tubes of energy stretched across the entire length of the ever-expanding Universe. These thin regions, left over from the early Cosmos, are predicted to contain huge amounts of mass and therefore could warp the space-time around them.

Cosmic strings are either infinite or they're in loops, with no ends, scientists say. The approach of two such strings parallel to each other would bend space-time so vigorously and in such a particular configuration that might make time travel possible, in theory.

Time machines

It is generally understood that travelling forward or back in time would require a device, a time machine, to take you there. Time machine research often involves bending space-time so far, that time lines turn back on themselves to form a loop, technically known as a "closed time-like curve."

To accomplish this, time machines often are thought to need an exotic form of matter with so-called "negative energy density." Such exotic matter has bizarre properties, including moving in the opposite direction of normal matter when pushed. Such matter could theoretically exist, but if it did, it might be present only in quantities too small for the construction of a time machine.

However, time-travel research suggests time machines are possible without exotic matter. The work begins with a doughnut-shaped hole enveloped within a sphere of normal matter. Inside this doughnut-shaped vacuum, space-time could get bent upon itself using focused gravitational fields to form a closed time-like curve. To go back in time, a traveller would race around inside the doughnut, going further back into the past with each lap. This theory has a number of obstacles, however. The gravitational fields required to make such a closed time-like curve would have to be very strong, and manipulating them would have to be very precise.

Grandfather paradox

Besides the physics problems, time travel may also come with some unique situations. A classic example is the grandfather paradox, in which a time traveller goes back and kills his parents or his grandfather, the major plot line in the "Terminator" movies or otherwise interferes in their relationship. Think "Back to the Future", so that he is never born or his life is forever altered.

If that were to happen, some physicists say, you would not be born in one parallel Universe but still born in another. Others say that the photons that make up light prefer self-consistency in timelines, which would interfere with your evil, suicidal plan.

Some scientists disagree with the options mentioned above and say time travel is impossible no matter what your method be.

Also, humans may not be able to withstand time travel at all. Travelling nearly the speed of light would only take a centrifuge, but that would be lethal.Using gravity would also be deadly. To experience time dilation, one could stand on a neutron star, but the forces a person would experience would rip himapart first.

So, is time travel possible?

While time travel does not appear possible, at least possible in the sense that the humans would survive it, with the physics that we use today. But this field is constantly changing, so keep tracking this field! Advances in quantum theories could perhaps provide some understanding of how to overcome time travel paradoxes.

One possibility, although it would not necessarily lead to time travel, is solving the mystery of how certain particles can communicate instantaneously with each other, faster than the speed of light.

How quantum physics is changing the perception of Universe?

Quantum mechanics is the branch of physics that relates to the very small. It results in what may appear to be some very strange conclusions about the physical world. At the scale of atoms and electrons, many of the equations of classical mechanics, which describe how things move at everyday sizes and speeds, cease to be useful. In classical mechanics, objects exist in a specific place at a specific time. However, in quantum mechanics, objects instead exist in a haze of probability; they have a certain chance of being at point A, another chance of being at point B and so on.

In the 19th and 20th century the Universe was mainly explained around classical laws outlined by Newton and Einstein. Classical mechanics explains the laws of big bodies. Quantum mechanics bridges the gap of world of small (microscopic). Slowly and steadily advancement in quantum world is revealing the description of universe at its smallest level. Scientists are finding strong connections between reality and its smallest constituents and particle level. The history of development of Quantum Theory below describes how and in which direction quantum world is moving.

- In 1900, Planck made the assumption that energy was made of individual units, or quanta.

- In 1905, Albert Einstein theorized that not just the energy, but the radiation itself was quantized in the same manner.

- In 1924, Louis de Broglie proposed that there is no fundamental difference in the makeup and behaviour of energy and matter; on the atomic and subatomic level either may behave as if made of either particles or waves. This theory became known as the principle of wave-particle duality: elementary particles of both energy and matter behave, depending on the conditions, like either particles or waves.

- In 1927, Werner Heisenberg proposed that precise, simultaneous measurement of two complementary values, such as the position and momentum of a subatomic particles impossible. Contrary to the principles of classical physics, their simultaneous measurement is inescapably flawed; the more precisely one value is measured, the more flawed will be the measurement of the other value. This theory became known as the uncertainty principle, which prompted Albert Einstein's famous comment, "God does not play dice."

Although scientists throughout the past century have balked at the implications of quantum theory including Planck and Einstein, the theory's principles have repeatedly been supported by experimentation, even when the scientists were trying to disprove them. Quantum theory and Einstein's theory of relativity form the basis for modern physics. The principles of quantum physics are being applied in an increasing number of areas, including quantum optics, quantum chemistry, quantum computing, and quantum cryptography. Even quantum laws seem to be playing a role in the working of the nature. Recent development in quantum biology is revealing quantum design in living beings too. Theorists and philosophers have even explored to connect continuous and brain functioning around quantum principles.

I also see deep connection between quantum principles and Universe around us. We can find clear answers if start looking up and macro level along with looking deep in the quantum world.

Look Up Question 46
How long Universe will last?

There is no one answer to this question. There are many theories related to end of the Universe, which we will see in next questions and each has different explanation to this question. What I believe is what I have already mentioned in my quote "Infinity is the destination of the expanding Universe".

<h1 style="text-align:center">Look Up Question 47</h1>
What are different theories that explain the end of the Universe?

There are hundreds of known cosmic events that could eliminate the life on our planet. High energy solar flares, cataclysmic impact events, gamma ray bursts, a nearby supernova, and on and on. However, the chances of Earth suffering a life-ending global catastrophe are rather slim.

The end of everything is a little bit more difficult to predict, but that hasn't stopped scientists from speculating and theorising. There are four key theories that explain the end of Universe:

The Big Rip

Our Universe is expanding. It happens because of a mysterious form of energy known as "dark energy." We don't know what it is but, each year this dark energy causes the rate of expansion to increase.

One theory of our ultimate end relies on the assumption that this expansion will continue indefinitely until the galaxies, stars, planets, and matter (even the subatomic building blocks that comprise all matter) can no longer hold themselves together. At which point, they rip apart. This theory is called the Big Rip.

Whether or not this will happen depends on critical density (the boundary value between open models that expand forever and closed models that re-collapse).

According to Robert Caldwell, a theoretical physicist, if the Big Rip won out over all of the apocalyptic scenarios put forth in this piece, the event would occur in some 22 billion years, when the Sun has already transitioned from a main-sequence star to a red-giant (incinerating Earth as a result) and then into a white dwarf. If Earth did manage to survive intact, the planet would explode about 30 minutes before the grand finale.

The Big Freeze

Another popular scenario for the end of the Universe that relies on deciphering the true nature of dark energy is the Big Freeze (also referred to as Heat Death or the Big Chill).

In this scenario, the Universe continues to expand at an ever-increasing speed. As this happens, the heat is dispersed throughout the space while galaxies, stars, and planets are all pulled farther and farther from one another. In the very distant future, intelligent civilizations will look into the sky and think they are alone. Everything will be so far away that the light from distant stars and galaxies can never reach them. Eventually, planets, stars, and galaxies would be pulled so far apart that the stars would eventually lose access to raw material needed for star formation and thus, the lights would inevitably go out for good. It will continue to get colder and colder until the temperature throughout the Universe reaches absolute zero. At absolute zero, all movement stops. Nothing can exist in such a place, as there is literally no energy at all anywhere. This is the point at which the Universe would reach a maximum state of entropy. Instead of fiery cradles, galaxies would become coffins filled with the remnants of dead stars.

Many scientists (astronomers and physicists alike) believe this is one of the most probable scenarios.

The Big Crunch

The Big Crunch is thought to be the direct consequence of the Big Bang. In this model, the expansion of the Universe doesn't continue forever.

After an undetermined amount of time (possibly trillions of years), if the average density of the Universe was enough to stop the expansion, the Universe would begin the process of collapsing in on itself. Eventually, all the matter and particles in existence would be pulled together into a super dense state (perhaps even into a black hole-like singularity).

Such an event might have already happened before. Some scientists have theorized that the Universe we see is the result of a cyclic repetition of the Big Bang, where the first cosmological event came about after the collapse of a previous Universe. This is something called as conformal cyclic cosmology.

Unlike the first two scenarios, this model relies on the geometry of the Universe being "closed" (like the surface of a sphere). An event like this would be like a single breath: The Universe would "breathe out" the Big Bang, and "breathe in" the Big Crunch.

This could also happen if we see a reversal of dark energy's current expansion effect.

Similar to this theory (and the Big Bang) is the Big Bounce. A sort of symmetry is proposed here: The Universe is in a continuous cycle of expanding out and then collapsing onto itself. Effectively, we could be one of many iterations of the Universe. Perhaps even more eerie to think about is the idea that maybe each time the Universe resets, it plays out the same way. Perhaps the you that is currently reading this article right now is just one you out of $10^{googleplex}$ versions that existed before.

Ultimately, the Universe may be like the mythical phoenix: its death is actually a fiery rebirth.

The Big Slurp

This theory surfaced just a few years ago, after revelations were released about the true nature of the Higgs Boson, which is the particle that plays a role in granting mass to elementary particles.

In this model, if the Higgs Boson particle weighs in at a certain mass, it could indicate that the vacuum of our Universe may be inherently unstable, perhaps existing in a perpetual "metastable' state, something that has been discussed at length many times before.

If this were the case, our Universe might experience a catastrophic event when a "bubble" from another alternate Universe appears in ours. If this bubble exists in a lower-energy state than our bubble, the Universe could be completely annihilated.

In essence, it would cause all the protons in all matter in our Universe to decay. By proxy, so would we. If that doesn't sound unpleasant enough, this sort of a vacuum meta-stability event could happen at virtually any moment, anywhere in our Universe. The bubble could pop over and start expanding at light-speed until it swallowed us entirely.

Look Up Question 48
Is our Universe in simulation?

If you, I and every person and thing in the cosmos were actually characters in some giant computer game, we would not necessarily know it. The idea that the universe is a simulation sounds more like the plot of "The Matrix," but it is also a legitimate scientific hypothesis.

The simulation hypothesis proposes that all of the reality, including the earth and the universe, is in fact an artificial simulation, most likely a computer simulation.

Many works of science fiction as well as some forecasts by serious technologists and futurologists predict that enormous amounts of computing power will be available in the future. Let us suppose for a moment that these predictions are correct. One thing that later generations might do with their super-powerful computers is run detailed simulations of their forebears or of people like their forebears. Because their computers would be so powerful, they could run a great many such simulations. Suppose that these simulated people are conscious (as they would be if the simulations were sufficiently fine-grained and if a certain quite widely accepted position in the philosophy of mind is correct). Then it could be the case that the vast majority of minds like ours do not belong to the original race but rather to people simulated by the advanced descendants of an original race. It is then possible to argue that, if this were the case, we would be rational to think that we are likely among the simulated minds rather than among the original biological ones. Therefore, if we don't think that we are currently living in a computer simulation, we are not entitled to believe that we will have descendants who will run lots of such simulations of their forebears.

Look Up Question 49
What is consciousness?

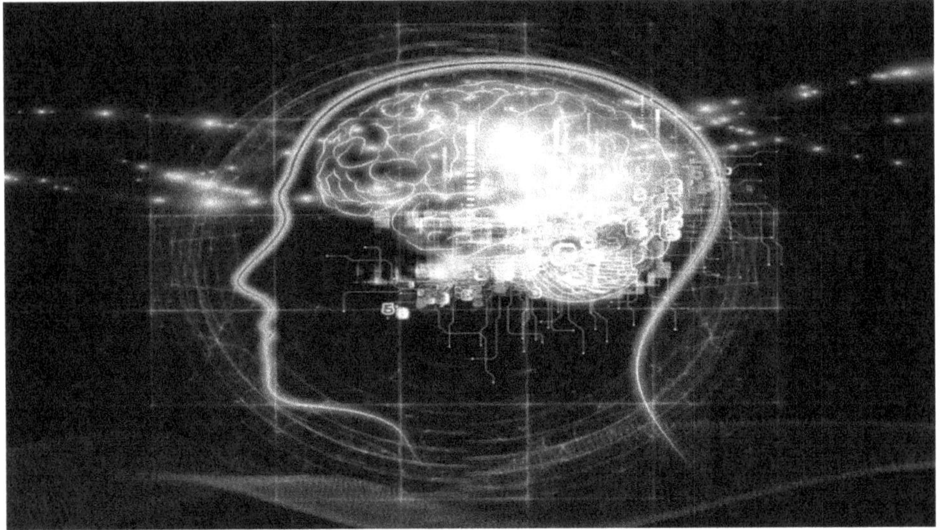

Figure: Does consciousness reside only in brain?

It's a metaphysical question that has preoccupied philosophers for centuries, but only recently have we developed the technology to address this question scientifically. Is consciousness something that emerges through the complex interactions of billions of cells? Is it a spectrum? Can it be replicated?

Consciousness is having of perceptions, thoughts, and feelings; awareness. The term is impossible to define except in terms that are unintelligible without a grasp of what consciousness means. Many falls into the trap of equating consciousness with self-consciousness, to be conscious it is only necessary to be aware of the external world. Consciousness is a fascinating but elusive phenomenon: it is impossible to specify what it is, what it does, or why it has evolved.

In medicine, consciousness is assessed by observing a patient's arousal and responsiveness, and can be seen as a continuum of states ranging from full alertness and comprehension, through disorientation, delirium, loss of meaningful communication, and finally loss of movement in response to painful stimuli.

We still don't know how to explain this in simple scientific way, but we are taking baby steps to ready the language of brain.

Look Up Question 50
Will Universe take birth again?

A cyclic model (or oscillating model) of Universe is any of several cosmological models in which the Universe follows infinite, or indefinite, self-sustaining cycles. For example, the oscillating Universe theory briefly considered by Albert Einstein in 1930 theorized a Universe following an eternal series of oscillations, each beginning with a big bang and ending with a big crunch; in the interim, the Universe would expand for a period of time before the gravitational attraction of matter causes it to collapse back in and undergo a bounce. So, with this theory Universe goes for endless cycles of birth and re-birth. The point to note here is this theory is yet to be approved.

Summary

In the preface of this book, I have narrated technically, what is Universe, what is life and how we are connected to the Universe. Out here in summary, I would like to use my philosophical taste to outline, what Universe tells us, what is the purpose of life in the Universe and why science is important to find out the connections between the Universe and the life we know on the Earth.

I strongly believe that our Universe has its own character. It operates with certain fixed laws, through which it governs the entire cosmos. It challenges us to explore and discover. It surprises us with its vastness. It makes us think that we are minuscule but still treats us as an important participant in the grand cosmic dance. It allows us to see its past by restricting the speed of light. It seems very calm to us from Earth, yet very vibrant at many other places. It has both the bright and the dark members in its family. It requests us to know him better through the language of mathematics and science. It keeps reminding us that everyone including us and him has a purpose!

I come under the bucket who believes that Universe is full of life, though we have yet to find. All types of life we see around is like a cosmic magic for me. How organic matter first converted into single cellular organisms. How multi-cellular organisms came into existence from primitive ones. How complex intelligent life like us evolved. Though we still don't know how life started on Earth, but I believe we are very close to find out if Universe is full of life. Any type of life, anywhere in the Universe, will have reflection of the key ingredients of the place where it evolved. We may not find life very similar to us out here on Earth. I will not be surprised if we find non-carbon, non-water based life somewhere else in the Universe.

The key similarity we will see is that every type of life will have strong dependence with the eco system they would be living in. In other words, there will be many cosmic connections we will have in the entire Universe with the life wherever we will find in future. We should use the language of Science and

Mathematics to explore these cosmic connections. This may give us some clues to answer some profound questions, like are we alone?

We have made our world dependent on technology all around us. Every part of our life is now getting benefits of advancements in technology. Wherever we go, we have gadgets, phone, internet, machines etc. to make our life better. But we have not invested enough in educating everyone to appreciate it with basic understanding of Science. Though in India, I must say I see strong inclination towards Science, but still at our planet Earth level, our species need to be more informed. With this book, I would like to promote the importance of Science in our daily life and understanding of the Universe with its language of Science and Math. Science and Maths are two fundamental building blocks of objective thinking. In schools, we should teach how to think rather than what to think. In every school, we should include curriculum on knowing about Universe. At home, we should motivate young kids, family members, elders and friends to read about Cosmos and often look up in the sky.

I would like to end this book now with some lines, a first attempt from me as a poet!

"Life goes smooth, want a break, do Look Up!

Mind with loops, need a shake, do Look Up!

Wish some time with dream and play, do Look Up!

Focus your mind, with lots of doubt, do Look Up!

Dimensions of love; draw on a page, do Look Up!

Desire time, along with space, do Look Up!

For innovate and contemplate, do Look Up!

Source of wisdom, refer anytime, do Look Up!

Life and death; need replay, do Look Up!"

By Kamal Deep Dham

References

Let me list down in the end what all and from where all I have collected all the wisdom in this book.

1. Book "Cosmos A Personal Voyage" By Carl Sagan.

2. Book "Astrophysics for People in a Hurry" By Neil deGrasse Tyson.

3. Book "Demon Haunted World Science as a candle in a candle in the Dark" by Carl Sagan.

4. Book "The Universe in a Nutshell" by Stephen Hawking.

5. Cosmos Episodes, the old series narrated by Carl Sagan and its re incarnation, narrated by Neil deGrasse Tyson.

6. I must mention another important source of information, the You Tube videos on Universe and Science. I have gone through hundreds of them in form of lectures and debates. I personally like all the debates organised by World Science Festival body.

7. I did also refer lot of organised, unorganised content on the internet which comes in Google search.

8. Knowledge around cosmic connections mainly came from various Facebook groups on Universe, Science and Astrophysics.

"Infinity is the destination of the expanding Universe!'

By Kamal Deep Dham

www.ingramcontent.com/pod-product-compliance
Lightning Source LLC
LaVergne TN
LVHW080425200726
843507LV00004B/727